PERFORMANCE INDICATORS TO MONITOR, ASSESS AND IMPROVE KNOWLEDGE MANAGEMENT PROGRAMMES IN NUCLEAR ORGANIZATIONS

The following States are Members of the International Atomic Energy Agency:

AFGHANISTAN
ALBANIA
ALGERIA
ANGOLA
ANTIGUA AND BARBUDA
ARGENTINA
ARMENIA
AUSTRALIA
AUSTRIA
AZERBAIJAN
BAHAMAS
BAHRAIN
BANGLADESH
BARBADOS
BELARUS
BELGIUM
BELIZE
BENIN
BOLIVIA, PLURINATIONAL STATE OF
BOSNIA AND HERZEGOVINA
BOTSWANA
BRAZIL
BRUNEI DARUSSALAM
BULGARIA
BURKINA FASO
BURUNDI
CABO VERDE
CAMBODIA
CAMEROON
CANADA
CENTRAL AFRICAN REPUBLIC
CHAD
CHILE
CHINA
COLOMBIA
COMOROS
CONGO
COOK ISLANDS
COSTA RICA
CÔTE D'IVOIRE
CROATIA
CUBA
CYPRUS
CZECH REPUBLIC
DEMOCRATIC REPUBLIC OF THE CONGO
DENMARK
DJIBOUTI
DOMINICA
DOMINICAN REPUBLIC
ECUADOR
EGYPT
EL SALVADOR
ERITREA
ESTONIA
ESWATINI
ETHIOPIA
FIJI
FINLAND
FRANCE
GABON
GAMBIA, THE
GEORGIA
GERMANY
GHANA
GREECE
GRENADA
GUATEMALA
GUINEA
GUYANA
HAITI
HOLY SEE
HONDURAS
HUNGARY
ICELAND
INDIA
INDONESIA
IRAN, ISLAMIC REPUBLIC OF
IRAQ
IRELAND
ISRAEL
ITALY
JAMAICA
JAPAN
JORDAN
KAZAKHSTAN
KENYA
KOREA, REPUBLIC OF
KUWAIT
KYRGYZSTAN
LAO PEOPLE'S DEMOCRATIC REPUBLIC
LATVIA
LEBANON
LESOTHO
LIBERIA
LIBYA
LIECHTENSTEIN
LITHUANIA
LUXEMBOURG
MADAGASCAR
MALAWI
MALAYSIA
MALI
MALTA
MARSHALL ISLANDS
MAURITANIA
MAURITIUS
MEXICO
MONACO
MONGOLIA
MONTENEGRO
MOROCCO
MOZAMBIQUE
MYANMAR
NAMIBIA
NEPAL
NETHERLANDS, KINGDOM OF THE
NEW ZEALAND
NICARAGUA
NIGER
NIGERIA
NORTH MACEDONIA
NORWAY
OMAN
PAKISTAN
PALAU
PANAMA
PAPUA NEW GUINEA
PARAGUAY
PERU
PHILIPPINES
POLAND
PORTUGAL
QATAR
REPUBLIC OF MOLDOVA
ROMANIA
RUSSIAN FEDERATION
RWANDA
SAINT KITTS AND NEVIS
SAINT LUCIA
SAINT VINCENT AND THE GRENADINES
SAMOA
SAN MARINO
SAUDI ARABIA
SENEGAL
SERBIA
SEYCHELLES
SIERRA LEONE
SINGAPORE
SLOVAKIA
SLOVENIA
SOMALIA
SOUTH AFRICA
SPAIN
SRI LANKA
SUDAN
SWEDEN
SWITZERLAND
SYRIAN ARAB REPUBLIC
TAJIKISTAN
THAILAND
TOGO
TONGA
TRINIDAD AND TOBAGO
TUNISIA
TÜRKİYE
TURKMENISTAN
UGANDA
UKRAINE
UNITED ARAB EMIRATES
UNITED KINGDOM OF GREAT BRITAIN AND NORTHERN IRELAND
UNITED REPUBLIC OF TANZANIA
UNITED STATES OF AMERICA
URUGUAY
UZBEKISTAN
VANUATU
VENEZUELA, BOLIVARIAN REPUBLIC OF
VIET NAM
YEMEN
ZAMBIA
ZIMBABWE

The Agency's Statute was approved on 23 October 1956 by the Conference on the Statute of the IAEA held at United Nations Headquarters, New York; it entered into force on 29 July 1957. The Headquarters of the Agency are situated in Vienna. Its principal objective is "to accelerate and enlarge the contribution of atomic energy to peace, health and prosperity throughout the world".

IAEA-TECDOC-2080

PERFORMANCE INDICATORS TO MONITOR, ASSESS AND IMPROVE KNOWLEDGE MANAGEMENT PROGRAMMES IN NUCLEAR ORGANIZATIONS

INTERNATIONAL ATOMIC ENERGY AGENCY
VIENNA, 2025

For further information on this publication, please contact:

Nuclear Knowledge Management Section
International Atomic Energy Agency
Vienna International Centre
PO Box 100
1400 Vienna, Austria
Email: Official.Mail@iaea.org

Printed by the IAEA in Austria
March 2025
https://doi.org/10.61092/iaea.wqyh-4n7l

IAEA Library Cataloguing in Publication Data

Names: International Atomic Energy Agency.
Title: Performance indicators to monitor, assess and improve knowledge management programmes in nuclear organizations / International Atomic Energy Agency.
Description: Vienna : International Atomic Energy Agency, 2025. | Series: IAEA TECDOC series, ISSN 1011-4289 ; no. 2080 | Includes bibliographical references.
Identifiers: IAEAL 25-01736 | ISBN 978-92-0-104125-8 (paperback : alk. paper) | ISBN 978-92-0-104225-5 (pdf)
Subjects: LCSH: Knowledge management. | Nuclear facilities — Employees — Training of. | Nuclear power plant operators — Training of.

FOREWORD

The IAEA continues to work with Member States to draw conclusions and identify lessons from past experiences in the use of nuclear energy. This enables it to develop a better understanding of current and emerging challenges, to share solutions and to work collectively with Member States to ensure that the economic and humanitarian benefits of nuclear power can be achieved in a safe and sustainable manner.

The IAEA supports Member States in developing knowledge management programmes and initiatives by organizing and supporting schools, training events, educational networks and assist missions. It also develops publications to address issues and challenges relating to critical knowledge acquisition and preservation and to share methods and tools beneficial for initiating, sustaining and continuously improving knowledge management.

For the application, successful delivery and continuous improvement of knowledge management initiatives, indicators are needed to help understand and monitor the status, performance, trends, strengths and weaknesses of knowledge management programmes and processes. Such indicators support sustainable knowledge management in accordance with an organization's corporate goals and strategies.

This publication was developed to assist leaders, managers and other employees with roles and responsibilities in nuclear power organizations and facilities to develop performance indicators to monitor and assess the effectiveness of knowledge management programmes and processes. It presents information on good practices observed through IAEA missions and activities and the knowledge and experience of different Member States.

The IAEA expresses its appreciation to the many experts from Member States who contributed to the development and review of this publication. The IAEA officers responsible for this publication were H. Zhivitskaya and T. Bilic Zabric of the Division of Planning, Information and Knowledge Management.

CONTENTS

1. INTRODUCTION

1.1. BACKGROUND

Knowledge is continuously being created, obtained and shared within organisations, including nuclear power organizations, throughout their entire lifecycle. Knowledge and its management are key components for the continuous improvement of an organization and to achieve the required culture for nuclear safety and performance. As an illustration of this key aspect, the word "knowledge" is included 76 times in the 'World Association of Nuclear Operators' (WANO) Performance Objectives and Criteria [1], notably, in the performance areas of nuclear professionals, organizational effectiveness, performance improvement and operating experience.

Maintaining the effectiveness of a knowledge management programme entails regular monitoring, measuring and assessment of the knowledge management (KM) practices and programmes with the aim to improve performance using specific indicators that have been developed explicitly for KM [2]. These indicators can be classified into Key Performance Indicators (KPIs) and Performance Indicators (PIs), each serving a unique purpose in performance assessment.

Key Performance Indicators are high-level, strategic metrics specifically chosen to reflect the most critical aspects of performance in KM that are aligned with the organization's overall objectives and success. They offer a clear focus on priority areas and are usually limited in number to ensure they highlight significant trends or gaps in KM effectiveness.

On the other hand, PIs encompass a broader range of metrics that capture various aspects of KM practices and programmes. They offer detailed, operational-level insights that inform day-to-day management and operational adjustments but are not always directly tied to overarching organizational goals. The careful selection and use of relevant PIs and KPIs are essential to ensure the successful delivery and realization of the benefits of KM and the effectiveness of KM programmes. This publication includes experiences and techniques from Members States for measuring performance. It provides useful insights on how the techniques can be implemented for the continuous improvement of KM programmes in nuclear organizations.

1.2. OBJECTIVE

This publication provides specific methods and tools for the identification, development, selection, establishment and use of PIs to measure and report on the implementation and effectiveness of KM programmes to improve the overall performance of a nuclear organization.

While the guidance is intended primarily for the use by the senior and middle level management of the owners/operators of nuclear power plants (NPPs) and/or utilities, the concepts and approaches are useful for all the other nuclear organizations. It may also be useful for other relevant stakeholders, including technical and scientific support organizations, regulators and educational providers.

1.3. SCOPE

This publication covers approaches by various organizations in establishing key performance indicators (KPIs) and Performance Indicators (PIs) to be used for monitoring and measuring the performance of KM programmes and associated processes to ensure successful implementation and continuous improvement of knowledge management. KPIs are selected and used to measure high-value, critical aspects of the overall KM activities, while PIs may address a broader range of KM processes to provide detailed feedback on the status and effectiveness of KM programmes.

This publication presents examples that are applicable to the entire lifecycle of a nuclear organization, including the decision, design, construction, commissioning, operation (including life extension), modernization and decommissioning phases. It can be used by organizations that have already implemented a KM programme or some of its elements. Organizations starting the implementation of a KM programme may use this publication and the examples of PIs, as a set of questions to indicate what need to be measured in the different phases of the programme.

1.4. STRUCTURE

The publication consists of six Sections including the introduction in Section 1 and the conclusions in Section 6, one Appendix containing a table with examples of KPIs and six Annexes with examples of good practices of KM KPIs establishment and use. Section 2 through Section 4 answer the questions 'why', 'what' and 'how' for the measurement of the performance of KM, KM programme activities and associated processes through a systematic and holistic KPI system, respectively, as follows:

— Section 2 explains the requirement and purpose for measuring the effectiveness of KM and KM programmes through a KPI;
— Section 3 discusses the construction and structuring of the KM KPI framework by linking the foundation, i.e. strategic business objectives and global performance areas, to specific KM and KM programme performance areas and criteria that are to be measured by the KM KPIs;
— Section 4 describes a KM KPI determination process to establish the ways to measure performance;
— Section 5 discusses the use of KPIs for measuring the effectiveness of KM practices and associated programme implementation for all stages of a nuclear project/plant lifecycle, from project preparation to decommissioning;
— Appendix provides examples of KPIs in eight functional areas and different life phases of a nuclear facility.

2. REQUIREMENT, PURPOSE OF MEASURING THE EFFECTIVENESS OF KNOWLEDGE MANAGEMENT AND PRINCIPLES FOR DEVELOPING KPIS

It is important to establish clear goals and objectives for any KM programme. These goals need to be specific, measurable, achievable, relevant, and time-bound (SMART). Once the goals have been established, appropriate metrics can be identified to track progress towards their achievement. It is also important to regularly evaluate the KM programme and adjust as necessary to ensure that it continues to meet the needs of the organization.

2.1. REQUIREMENT FOR MEASURING THE EFFECTIVENESS OF A KNOWLEDGE MANAGEMENT PROGRAMME

Knowledge is not an isolated commodity, and its management is not a simple stand-alone task. Knowledge is a primary asset and one of the key resources of any business or institution, with corporate knowledge being an integral part of business operations contained within processes, people, technology, and tools that need to be managed. Therefore, proactively managing and positively influencing and harnessing the knowledge that is important to the organization's purpose or business makes a difference in the organization's commercial success and the performance of all (current and future) activities to achieve its goals. A KM programme that is integrated within all the relevant programmes and processes helps in the effective management of the knowledge. Thus, the KM provides an effective management of the intangible assets of organization which creates its business value, the monitoring and assessment of the KM programme implementation becomes critical for each organization.

The most common reasons for measuring the effectiveness of KM programme are to monitor performance improvement as well as the progress and success of a particular KM initiative. The most important consideration in the selection of PIs is to ensure that there is a clear link to the organizational strategy and objectives, and that they cover the main areas that will benefit from the KM initiative. Details on the incorporation of PIs in the overall management system and support of the organization's business case, as well as their use for continuous performance improvement, are provided in Annex I.

In many cases, the overall organizational PIs, such as those developed as a part of an integrated management system process, did not fully or clearly provide an indication of the decline, or the attention required, of the KM programme until a lack of knowledge hindered a process or an activity. Even then, the measures were not able to reveal how knowledge, or lack of it, prevented the process from reaching the expected performance. Section 4 explains the difference between PIs and KPIs.

Only an integrated and balanced set of PIs can provide a meaningful and comprehensive evaluation of the effectiveness of the processes and subprocesses, which can be improved through more systematic work in the areas of knowledge management.

2.2. PURPOSE OF MEASURING THE EFFECTIVENESS OF KNOWLEDGE MANAGEMENT

The development and use of appropriate PIs helps leaders and managers, as well as other employees, to understand the effectiveness of the implementation and execution of KM programmes in optimising knowledge in an organization to better serve its business and strategic objectives and goals.

Setting up a system of PIs for KM to monitor and ensure that institutional knowledge remains at the expected level has two main objectives:

- To provide simple and direct information about the level of success or shortcomings of KM and its required, expected and desired benefits;
- To identify areas of both strengths and weaknesses that would provide direction for developing reliable actions, plans and forecasts for improving organizational tasks and behaviours to better achieve the expected benefits.

Accordingly, measuring the effectiveness of KM by PIs provides the status and trends of the acquisition, preservation and use of the required knowledge for current and future activities, processes and resources to:

- Align knowledge and KM elements, including the roles and responsibilities within the KM programmes of an organization, to daily actions and long-term plans;
- Proactively take action to retain or replace knowledge before the lack of knowledge hinders required and expected performance;
- Communicate and encourage behaviour to request, obtain and utilize the necessary critical knowledge in activity performance;
- Review and assess the links between knowledge and the critical success factors of an organization and, if necessary, adjust them in accordance with the desired organizational benefits and standard of performance;
- Encourage the correct behaviour that maintains the organizational performance and level of safety to deliver the desired results.

Knowledge management PIs provide the organization's decision makers and senior management with systematic information to understand the adequacy and appropriateness of the KM tools, methods and procedures within the framework of the organization's business model. Accordingly, measuring the effectiveness of any KM programme considers the performance of three primary elements, people, processes and tools, of a KM system, towards achieving the overall organizational performance objectives.

2.3. PRINCIPLES FOR DEVELOPING, USING AND MANAGING KEY PERFORMANCE INDICATORS

This publication includes many good practices and examples that nuclear organizations worldwide have found beneficial. In addition to these good practices, the following five main principles are considered to be the most important success factors when developing, using and managing KPIs:

1. The goals of a KM KPI system need to be clearly stated and aligned to support the achievement of the overall business objectives.
2. The responsibilities, requirements and supporting arrangements for the development, use and oversight of KM KPIs need to be integrated within the organizational business process and management systems.
3. A combination of indicators, including short and long term, leading (i.e. output orientated) and lagging (i.e. input orientated) need to be used to measure KM performance.
4. PIs need to measure the relevant KM attributes in a clear, timely and efficient manner, with the analysis clearly and consistently reported and communicated.
5. Arrangements need to be in place to enable improvements of the KM KPI system.

2.3.1. Principle 1

Introducing KM initiatives involves considerable time and resources of an organization. The contribution that effective KM programmes can make to safety and business objectives need to be clearly identified when setting up the programmes. Identifying the KPIs to monitor and evaluate the effectiveness of KM programmes is essential to substantiate achievements and can be used to support the delivery of the overall business objectives as well as improving business performance. This is supported though clearly stated goals which demonstrate a link to organizational objectives.

2.3.2. Principle 2

Knowledge and information systems are a resource of the organization and therefore, KM processes need to be embedded as part of the organizational business process and management system. The responsibility of these processes and elements needs to be clearly identified and outlined throughout the organization.

Internal and external stakeholders need to understand their roles and responsibilities in relation to achieving the KM goals, and the importance of effectively utilising KPIs. Clear oversight arrangements need to be outlined and effectively utilized.

2.3.3. Principle 3

To ensure that a KM programme is developed and maintained effectively, it is important that a minimum set of KPIs is produced and that they meet the users' needs. Whilst PI selection is inevitably influenced by the availability of data; it is important to note that there may be copious amounts of data available. It is therefore important to be selective and to ensure that the indicators are linked back to the business objectives.

In deciding on the KPIs, a variety of indicators need to be utilized to provide a rich and meaningful source of information for analysis. Both long- and short-term indicators need to be considered, along with leading indicators which provide information on output, and lagging indicators which focus on the input.

It is important to regularly review the choice of PIs to ensure they are still fit for purpose and provide the relevant information for stakeholders.

2.3.4. Principle 4

Performance indicators can be used to set and communicate performance goals and to identify possible gaps. Therefore, it necessitates clear and consistent communication of trends, analysis, and findings.

Defining, calculating, and reporting the PIs must be done clearly and concisely. The chosen indicators should be unambiguous and conveyed in a way that aids stakeholder understanding [3-8].

The target date of the indicators and any analysis need to be agreed upon and understood by all. The results should allow organizations to respond to performance deterioration in a timely manner.

2.3.5. Principle 5

A performance indicator system can be enhanced if it is subject to a review, analysis, and improvement process. This may result in choosing different indicators, amending the indicators, changing the way the indicators are calculated, or changing the way the information is analysed and presented [9].

These ongoing regular reviews, self-assessments, and benchmarking need to be performed with managerial evaluation assessing the validity and meaningfulness of the KPIs to ensure that they reflect the performance measurement needs, which in turn support existing strategies and business objectives.

3. SETTING UP A KNOWLEDGE MANAGEMENT KEY PERFORMANCE INDICATOR SYSTEM

In accordance with the roadmap defined in the IAEA's Guide to Knowledge Management Strategies and Approaches in Nuclear Organisations [2], introducing KM initiatives involves considerable time and resources of an organisation. The contribution that effective KM programmes can make to safety and business objectives needs to be clearly identified. Performance Indicators are essential to monitor and evaluate the effectiveness of KM programmes as well as to substantiate achievements. They also can be used to amend a programmatic activity to suit the business requirements.

The use of PIs provides a way to measure an activity to ensure its achievement. By looking at the KM implementation phases, three levels of KM maturity have been outlined:

— Level 1: KM awareness and orientation;
— Level 2: KM implementation and roll-out;
— Level 3: KM advanced implementation.

As it was indicated in the IAEA's Guide to Knowledge Management Strategies and Approaches in Nuclear Organizations [1] there are eight organizational or functional areas in the development and implementation of a KM programme.

Figure 1 illustrates the grouping of eight organizational or functional performance areas, which are typical for the initiation and operational stages of a KM programme with KM tools and activities under three main components: people, processes and tools. Details on the performance areas are provided in Annex II.

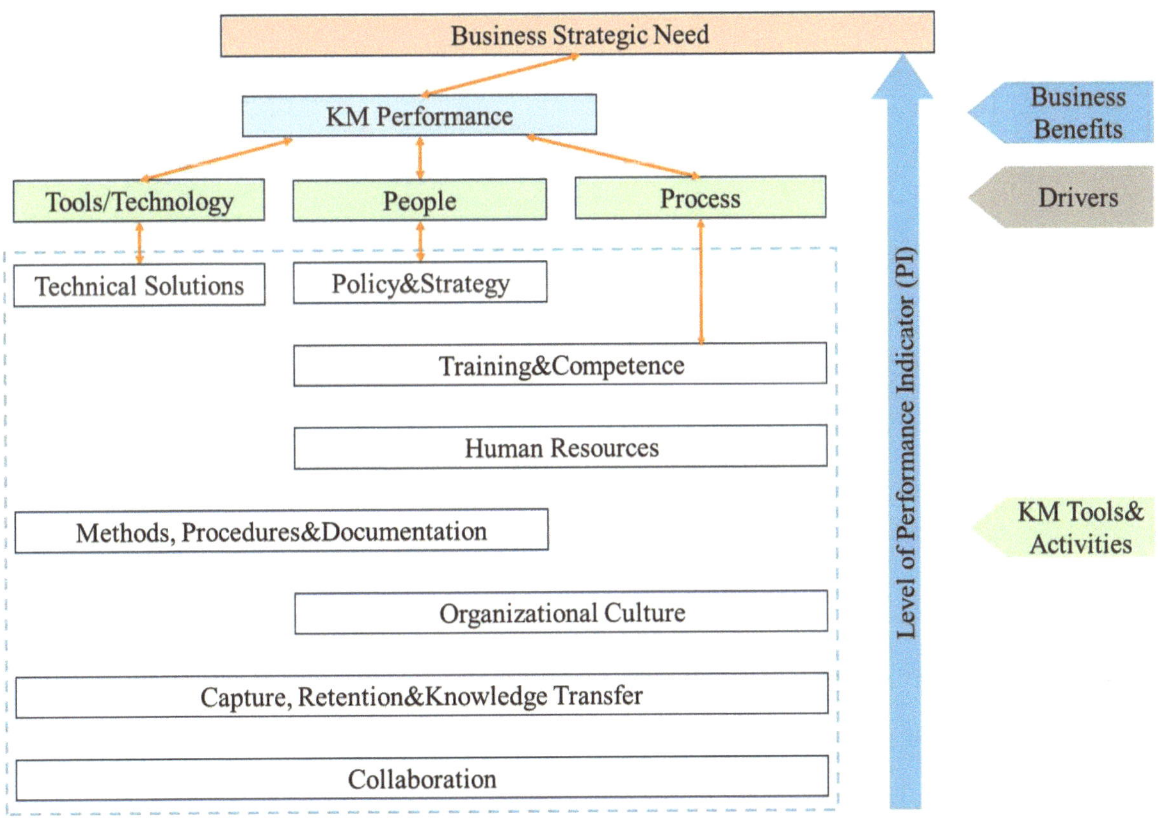

FIG. 1. An illustration of a KM PI framework.

Figure 2 shows the correlations between the maturity levels of an organization, the implementation phases and the relevant modes to measure the effectiveness of knowledge management.

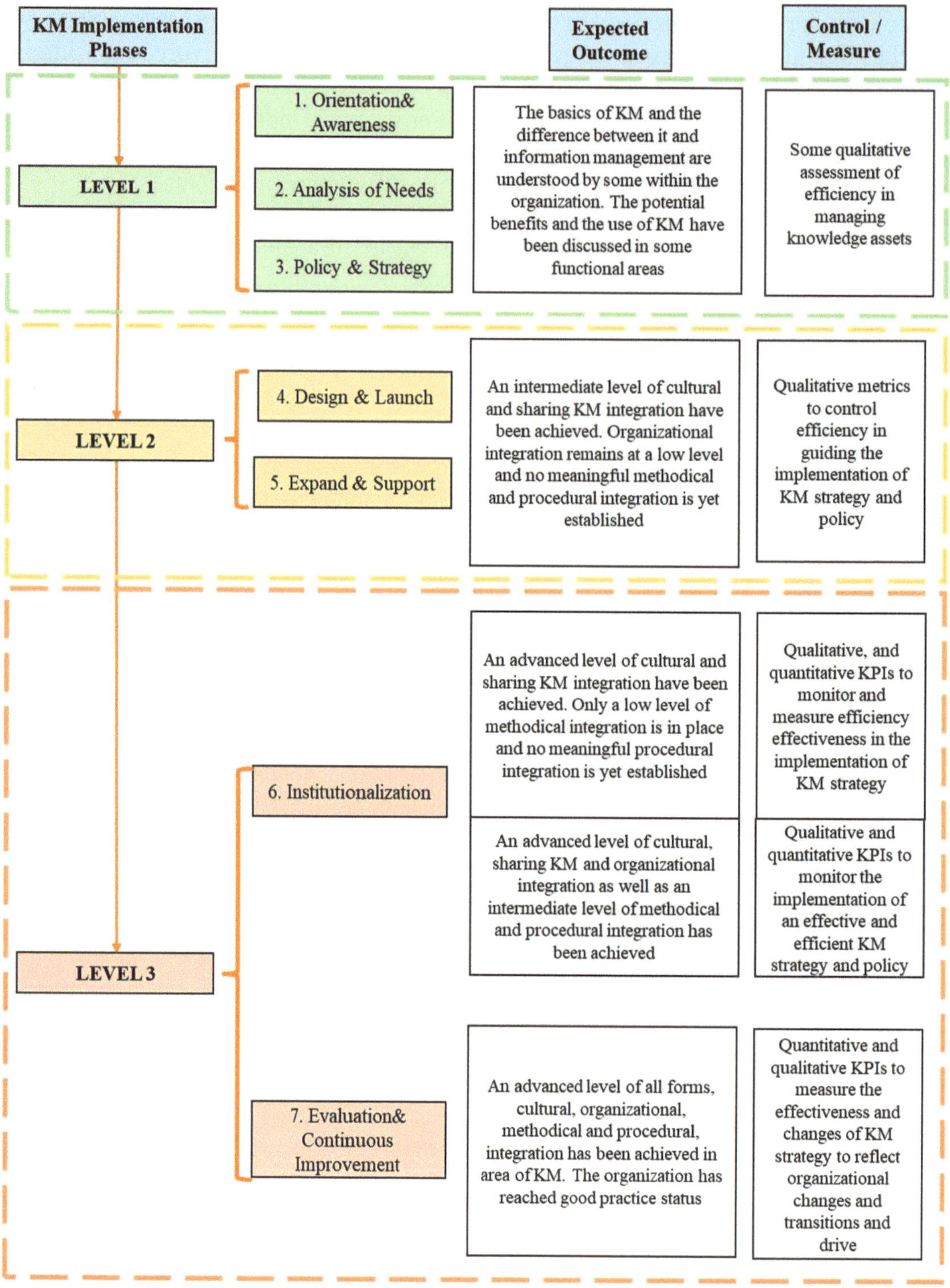

FIG. 2. A correlation of the implementation phases of a KM programme and the modes of measuring the effectiveness of KM.

The suggested framework for KM PIs considers the maturity levels of eight KM functional areas, as well as all the stages of the lifecycle of the nuclear organization.

The process of identifying appropriate PIs for each performance area starting from an organization's strategic objectives is described in the following sections. As the organization moves through the different stages, some specific PIs may only be relevant for some of the stages [10-20]. The need to continuously review and improve the set of PIs is defined as one of the principles in developing and improving PIs and is emphasized several times in this publication. Examples of PIs for different functional areas and different stages are provided in the Appendix.

4. ESTABLISHMENT OF KEY PERFORMANCE INDICATORS

Indicators to measure the performance status and progress of KM programmes, processes, and practices need to be aligned with the organization's business profile, its strategic goals, and its critical success factors [21-27].

The process of establishing KPIs for KM considers three fundamental questions in selecting specific indicators:

— What can be measured to indicate the level of achievement of a specific KM goal?
— How can an indicator be set and assessed to objectively provide information on the current status and where improvements can be made in achieving a specific KM goal?
— How to identify the optimal number of KM PIs considering PIs developed within other management systems and the need for continuous maintenance, improvement/correction of PIs?

Management systems integrate KM programmes with other programmes within an organisation, it may therefore be possible to measure some KM PIs within the integrated management systems as well as the KM programme. A system to determine a smaller set of high-value PIs representing a substantial portion of the characteristics of the past, current, and future status and performance of KM, with a low impact on resources can be achieved by paying attention to both:

— The starting point, 'to serve/support specific business strategic objectives', which is defined by the established business strategic objectives and associated overarching global standards of performance; and
— The end result, 'what we are going to do with the KM indication to support the business strategic objectives', which is defined by the strategy, philosophy, values and expectations for the institutional knowledge — to best serve the business strategic objectives — that are declared in the KM policy and strategy.

4.1. LAYERING THE STANDARDS OF PERFORMANCE AND CRITERIA AT APPROPRIATELY APPLICABLE LEVELS

Achievement of business objectives and the goals of global performance areas are vital for the purpose of business. The global performance areas corresponding to the strategic objectives define, focus and align performance of daily functions and activities in all areas throughout the organisation, as well as plans for future ones, towards the purposed outcome of the business. Typically, the strategic objectives and associated global performance areas related to use, maintenance and protection of resources (e.g., physical, financial and human assets, business processes, safety, know-how) or progress of outcomes (such as growth, innovation, customer satisfaction, learning/knowledge, etc.). However, in nuclear area, some performance areas and categories are unique and differ from most other sectors.

In accordance with the specific business objectives and performance categories, the organisations establish their global standards of performance to describe the requirements and expectations of performance in every activity performed by every person in the organisation in every operational area. Therefore, primary purposes of the standards of performance [28-31] are to:

— Set performance expectations for the identification, assessment, and management of sustainability-related risks for continuously and successfully achieving business objectives, as well as for the identification of opportunities for planning to maintain success and sustainability;

— Enable business functions and assets to design and operate systems, programmes, and processes with assumed or delegated role and responsibility (and associated accountability);
— Provide criteria for measuring performance as to value creation and business improvement (that are measurable and unambiguous, which is the key purpose of establishing a PI selection and utilization system).

When building the organization's PIs in a systematic way, it is necessary to consider that the PIs will support decisions at all levels:

— Strategically, at the authorization level;
— Tactically at the administration level;
— Programmatically at the programme and process coordination level;
— Operationally (task-specifically) at the activity execution level.

Global standards of performance and associated performance criteria, and the corresponding PIs at the strategic level, are decomposed into tactical and operational levels as they mandate the specific standards of performance for programmes, processes, and tasks, including those related to and relevant to KM programmes. A top-down approach for performance requirements and expectations with the formation of strategic-level performance criteria (and associated PIs) can be separated into a number of tactical-level performance criteria (and PIs) in all the specific performance areas of an organization. Consequently, the tactical-level performance criteria (and PIs) establish the operational-level performance criteria and PIs in all activities of the organization. This will support the achievement of the specific performance of programmes, processes, organizations, and tasks — including those that are for (or relevant to) KM — that achieve the business strategic objectives in a bottom-up manner, as depicted in Fig. 3.

It needs to be noted that, theoretically, there is almost no limit to the number of PIs that can be established in an organization for different purposes for the measurement of the performance of programmes, processes, organizations, activities, tasks, individuals, etc. Therefore, the number of PIs needs to be optimized and weighted in accordance with their significance and necessity. While the PIs will be determined for improvement of performance within each level, those that are the most significant for the overall performance at every level will be particularly monitored, evaluated, and communicated from bottom-to-top. Conversely, the actions that are determined from the results of their assessments will be driven at every level from top-to-bottom, as shown in Fig. 4.

Often, when establishing performance measurement systems, there is a challenge in defining the PIs versus the KPIs i.e., those that are focusing on aspects of organizational performance that are the most critical for the current and future success of the organization. Both PIs and KPIs are metrics that can be used to measure performance and provide information on the current level of performance status and progress. However, they differ in their impacts on the overall strategic business objectives and relevancy to the audiences (users) within the organization and among the stakeholders and the extent of decision making based on their indications and assessments. More specifically, KPIs enhance the understanding of the strategic, tactical, and operational performance of programmes, processes, and resources to identify strengths and weaknesses and enable further improvements, including those of KM and KM programmes.

To determine and establish the KM PIs and KPIs at each level, it is necessary to differentiate the global and specific standards of performance and performance criteria into strategic, programmatic, tactical, and operational level KM performance criteria and KPIs according to their KM programme attributes.

An organization can use the same, similar or different KPIs linked to strategic, programmatic, tactical or operational objectives and performance expectations, which are reviewed and acted upon at any level. Some of these KPIs will directly measure the performance of KM programmes and KM processes and activities/tasks, while some will measure the KM related performance of other programmes, processes, and activities.

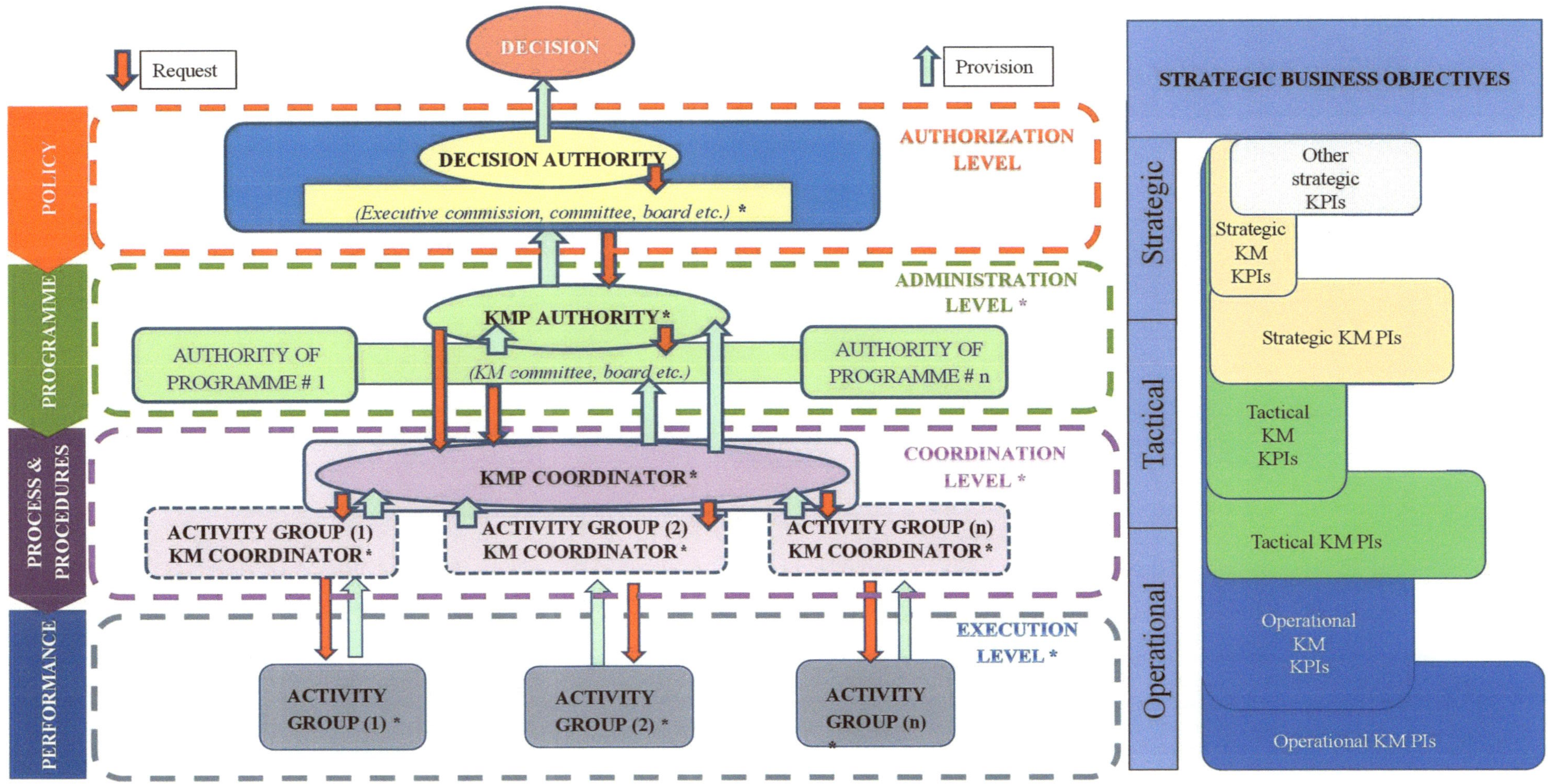

FIG. 3Decision levels in knowledge management (KM), KM programme strategy and implementation, along with the establishment and communication of performance indicators (PIs) and key performance indicators (KPIs).

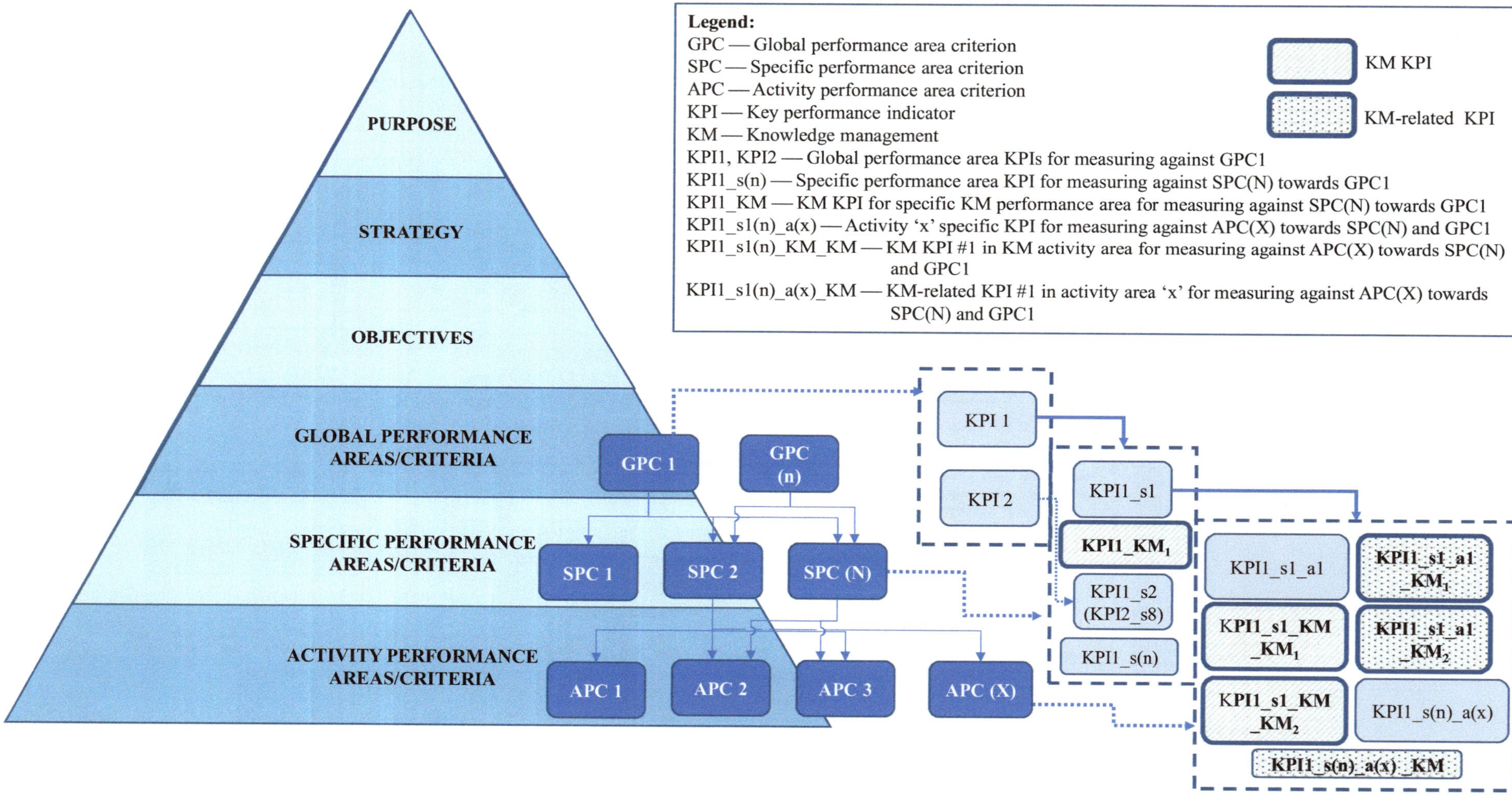

FIG. 4. Illustration of layering performance criteria and corresponding KPIs in determination of KM KPIs and KM-related KPIs.

An overall approach for identifying KM KPIs and KM-related KPIs inside an organization's performance assessment framework is illustrated in Figures 3 and 4. Several examples are provided in Annex III.

The examples of PIs provided in the Appendix, are not grouped in the four levels, as described above. They may belong to different levels depending on organization goals, maturity level, and management of knowledge. However, the approach explained above, the example presented below and the examples in Annex III provide information on how qualifying PIs in different organizational levels depend on their specific cases.

Example:

Strategy:	Competent staff in a nuclear organization
Strategic Objective:	Knowledge Sharing Culture
Global performance area:	Knowledge Loss Risk Management
Specific performance area:	Transfer of knowledge – Competency building
Activity performance area:	Exit interviews/identification of critical knowledge – Corporate knowledge building

Knowledge Loss Risk and Transfer of Knowledge is a typical KM initiative that many nuclear organizations are applying. It involves an initial risk analysis to identify staff/positions holding critical knowledge, followed by different techniques/approaches to transfer/capture knowledge. For this initiative, the following KPIs, which can be found in the Appendix, are proposed:

The following KPIs are for the operational level (task):

— Number of critical knowledge holders at risk;
— Number of exit interviews for knowledge transfer with the critical knowledge holders;
— Number of knowledge transfer activities undertaken versus planned;
— Number of successors identified to receive knowledge and skills from critical knowledge holders;
— The amount of documented knowledge produced from knowledge captured from critical knowledge holders.

The following KPIs are for a programmatic level:

— Percentage of realized knowledge transfer (any KM tool used) compared to the number of knowledge holders who left in the calendar year;
— Number of situations that caused delay/cancellation/failure/errors in jobs or activities due to the lack of critical knowledge holders;
— Turnover rate of in-house and transferred knowledgeable staff.

The following KPI is for a tactical level:

— Effective mechanisms are in place to promptly transfer an organization's values and expectations regarding the transfer of knowledge to new staff (qualitative YES/NO).

The following KPI is for a strategic level:

— The organization's values and behaviours for the transfer of knowledge are modelled by its leaders and practiced by all plant staff (qualitative YES/NO).

4.2. GENERIC METHODOLOGY FOR DEVELOPING, USING AND MANAGING PERFORMANCE INDICATORS

This section provides a generic methodology for developing, implementing, and managing PIs and a guide on how to manage the process of developing and managing PIs in a structured manner. It is, however, not intended to be taken literally or as a required standard, but as the result of good practice.

The KM PI system will be built step-by-step:

1) Acknowledgement of the strategic business objectives and global standards of performance with regard to knowledge;
2) Acknowledgement of the corporate strategy for the critical knowledge that is determined to be essential for specific strategic business objectives and global standards of performance;
3) Acknowledgment of a KM strategy and KM programme goals to ensure accomplishment of the business strategic objectives in accordance with the corporate strategy and policy for managing knowledge;
4) Acknowledgment of KM standards of performance and performance criteria;
5) In accordance with the programme goals, set KM standards of performance for:
    - Each performance area;
    - Each knowledge lifecycle aspect, i.e. knowledge identification, acquisition, classification/selection, registration/preservation, distribution/dissemination, utilization, modernization, and transmission in accordance with the KM programme goals;
    - Each KM element, i.e. people, technologies/tools and processes;
6) Place the KM standards of performance into strategic, programmatic, tactical, and procedural (task-specific) levels according to their attributes;
7) Determine the weighting of each KM standard of performance by assigning a value of its significance in contributing to the overall requirements and expectations from the highest significance to lowest;
8) Establish PIs for each KM standard of performance which needs to be measured and assessed at all levels, and checking all the PIs that have been previously set in the organizations, programmes, processes, and procedures to:
    - Avoid duplications (i.e. an existing PI is used for the same measurement purpose as the new PI for a KM programme);
    - Substitute (i.e. an existing PI used for a different purpose can be modified to be used for a KM purpose instead of a new PI);
9) Set metrics, rules, methods, thresholds, and assessment criteria for each PI, and where known or beneficial, include a set of possible actions to be taken at all levels based on the outcome of the assessment of each indicator.
10) Set a feedback loop for the process, aligning the PI to the corporate strategy and values and expectations for knowledge to repeat the process for correction or fine tuning.

Once the KM related performance criteria are identified at each level and are affiliated with the KM components and scope, the next step is to select and establish proper KM PIs that would have the best alignment with the performance criteria that monitor and measure the success and improvement of KM performance.

Theoretically, there is almost no limit to the number of PIs that can be established in an organization for the measurement of the performance of programmes, processes, organizations, activities, tasks, individuals, etc. Therefore, the number of PIs needs to be optimized and weighted in accordance with their significance and necessity, considering that:

— A large number of PIs could provide more information on performance measurement and monitoring but would increase the cost (financial and time) of maintenance and management, particularly those associated with data gathering, review, and assessment, and would decrease the efficiency of the application and hinder communication;
— Not having a sufficient number of PIs will result in being totally or partially uninformed about the required or expected performance as to its status, trends, issues, and need for action for improvement or repair, as well as determining the appropriate actions to do so.

Therefore, in order to make monitoring and remedial action manageable and functional, the selection and establishment of the most informative, high-value PIs is essential to ensure satisfactory performance and improvement of programmes, processes, organizations, activities, tasks and individuals. The high-value PIs or KPIs, are indicators for decision making at different levels that are tightly connected to the strategic business objectives of the organization. Organizations themselves determine which PIs are linked with strategical, tactical, or operational objectives and which ones will be considered as KPIs or supporting indicators/metrics.

Detailed key considerations in selecting the type and number of PIs are provided in Annex IV.

4.3. ROLES AND RESPONSIBILITIES

The primary responsibility of a KM coordinator is the effective coordination of the knowledge management, and therefore the primary responsibility for the design and implementation of the PIs lies with the KM coordinator. The KM coordinator is responsible for developing new PIs and for changing or cancelling existing PIs. However, the KM coordinator needs to have knowledge of the entire PI process. The KM coordinator is responsible for overseeing and managing the collection, analysis, trending, and reporting of all the associated data and KPIs to senior management in accordance with internal procedures.

To avoid the development of too many PIs, which are not providing added value for measuring KM performance, it is recommended to establish an independent peer group for reviewing each request to develop new PIs and changing or cancelling existing PIs. An organization can establish a Performance Indicator Panel (Committee) with responsibility for providing oversight of the PIs to ensure that a comprehensive, consistent set of metrics is maintained across the organization. It can also ensure that an optimal number of PIs is maintained that balances the need to adequately monitor the performance of the organization with the objective to minimize the overall number of indicators. The organization's management is responsible for identifying and maintaining ownership of the PIs necessary to fulfil the organization's mission, to continuously monitor its performance, and interface with the PI Panel.

For effective implementation of a KM programme, every person in the organization is expected to know and understand their role and responsibilities.

5. KNOWLEDGE MANAGEMENT PERFORMANCE MEASUREMENT APPLICABLE TO DIFFERENT STAGES OF THE ORGANIZATION LIFECYCLE

The design and use of KM KPIs will clearly vary between organizations, depending on their strategic goals and objectives, but it will also vary according to the maturity level of an organization's KM programme, and the necessity of various KM initiatives during the lifetime of the nuclear organization. The majority of PIs may generally be the same in all stages and phases, what they are indicating and how significant they are in performance monitoring, measurement, and interpretation may be different, particularly with regard to the resulting actions from their assessments.

For the purpose of this publication, the KPI system for the measurement of KM performance describes the main functional areas of KM and the phases of the organization lifecycle by using a specific example of an NPP with a mature KM programme, and established KM practices.

The phases of an organization's lifecycle consist of:

— Project preparation, which consists of programme consideration and project planning and development processes and activities for all necessary decisions;
— Project/facility design and realization;
— Facility utilization;
— Facility decommissioning.

For each of the eight functional areas [2], in the Appendix provides examples of common KPIs as well as specific ones for each of the above four phases. The Appendix is a source of PIs which can be considered when establishing and improving systems of KM measurement. These PIs need to be reviewed together with Annex VI, which contains more information (description, threshold, frequency of reporting, etc.) for some PIs developed by different organizations and countries. As mentioned in Section 4, the PIs provided in the Appendix may be on an operational, programmatical, tactical, or strategical organizational level, such qualification of PIs is not provided in the Appendix.

5.1. PROJECT PREPARATION - KNOWLEDGE MANAGEMENT KEY PERFORMANCE INDICATORS

Project preparation and development cover the period of an NPP's lifetime from the consideration and affirmation of launching a nuclear programme for power generation with a new (or additional) NPP to getting ready to invite and evaluate bids for the NPP project (i.e. phases 1 and 2 as defined in Ref. [19]).

5.1.1. Pre-project Phase

The pre-project phase of an NPP's lifetime primarily includes the processes and activities that contribute to making a knowledgeable decision to launch a nuclear programme for power generation with a new (or additional) NPP and develop plans for a nuclear programme and its implementation.

Therefore, at this stage, a certain level of knowledge is needed to produce detailed evaluations and assessments and to clearly delineate the conditions and commitments inherent in undertaking a nuclear power programme. Some of this knowledge exists as permeated resources in a conglomerate of several organizations working together towards the decision for implementation of a nuclear power programme.

In a newcomer country, the KM will typically have two main aspects during the pre-project phase, which are described in more detail in Annex V:

— Identification and acquisition of the lifecycle elements of KM concerning primarily people/human and organization/process components;
— Specific KM performance areas focused on human resource development.

Examples of KPIs for monitoring and measuring the performance of KM activities in this phase, are provided in the Appendix.

5.1.2 Project Development Phase

The project preparation activities require a combination of technical, commercial, financial, and legal expertise with particular knowledge, for example, of electrical and nuclear technologies, safety and quality assurance, techno-, power- and environmental economics and law, as well as major project financing.

In a newcomer country, KM primarily focuses on the identification, acquisition, and preservation of knowledge related to the current and future strategic objectives of the decision-making entity, both for the overall programme/project and specifically for the management of knowledge.

An existing and experienced owner/operating organization planning a new (additional) power plant project is likely to already have a mature KM programme and will have established KM strategies and policies, commitments, and values, in addition to their own accumulated knowledge and operating experience or have access to other knowledge and experience in the nuclear and other industries [18-20]. An experienced organization undertaking a new (additional) power plant project likely already has a mature KM programme, established strategies, policies, commitments, and values, along with accumulated knowledge and operational experience, or access to expertise in nuclear and related industries. It needs to also be highlighted that, in some expanding nuclear programmes, the time period between a new project and the previous one may be long enough to detrimentally impact the design and construction knowledge and experience. In such situations, it may be necessary to refresh the knowledge by reviewing the latest experience of KM-related issues and events, the application of existing KM programme, potential gaps, pitfalls and lessons learned from other recent projects.

If the future NPP is to be maintained and operated by domestic resources, one of the key KM performance areas will be human resource development planning. Accordingly, there are common and specific performance criteria applicable to each stakeholder, such as academia, industry, government, or utilities.

In this phase, the knowledge and skills needed in the next phases as well as staff requirements, recruitment, and training plans based on capacity gap analyses of the involved organizations need to be identified. The plans usually cover competency, education/training, knowledge, and experience expectations and requirements based on the KM strategy of the decision maker and need to consider bilateral and international competence building activities for information and knowledge acquisition.

Examples of KPIs are provided in the Annex V.

5.2 PROJECT IMPLEMENTATION/PLANT REALIZATION KNOWLEDGE MANAGEMENT KEY PERFORMANCE INDICATORS

Once a project moves into the design and construction phase, the knowledge is located and shared by the organizations involved in the design, construction, and commissioning of the NPP.

The core expertise and knowledge that exist within a project owning organization enable the adequacy of plant design, construction, and commissioning to be assessed and verified to ensure that it meets or exceeds the owner's requirements, as well as overseeing and managing the implementation of the project.

Prior to the operational phase, the project owner needs to ensure that its organization preserves and enhances its existing institutional knowledge and information of systems, components, programmes, and procedures, since the personnel, together with their personal knowledge, will change throughout the phases of a nuclear programme, which in total could be of the order of a hundred years.

The initiation and application of an effective KM programme are important at this phase for the management of the institutional knowledge as the project owner prior to the operational phase will transfer all responsibility for the safe and reliable operation of the plant to the operating organization when the operating licence is acquired.

Typically, the KM programme will focus on the acquisition and preservation of knowledge, with some aspects of utilization and will include all three components of knowledge management, people, processes, and technology. Establishment of the programme needs to be considered by corporate management, with a key focus on the transfer, recording, and retention of knowledge.

This phase of KM will determine the future knowledge retained about the NPP for its safe, reliable, and economic operation, maintenance, and modification. Therefore, the high effectiveness of knowledge transfer (i.e. the acquisition and preservation of initial design and facility configuration knowledge) has to be monitored and its success measured by KPIs. Examples of such KPIs are provided in Annex V.

During the utilization phase, numerous decisions are required on plant activities and assets. The knowledge that is necessary to ensure that the correct decisions are taken needs to be available, appropriate, and sufficient, which requires an effective and efficient KM strategy with an established, mature, and structured KM programme. A key process to ensure the successful and continuous performance improvement of the KM programme is to measure and monitor its performance and all associated processes and procedures by established indicators that demonstrate the strengths or weaknesses.

Interfaces and interactions between organizations and between individuals occur continuously during plant operation, requiring departments and people to be collectively and continuously aware and vigilant of any possible issues related affecting institutional knowledge. An effective KM programme during operation therefore requires application of a systematic management process for all activities at the plant/site, e.g., operation, engineering, work planning, oversight, surveillance, testing, chemistry control, and radiological control. Other important knowledge areas are administration and coordination of the knowledge of the plant programmes and processes, which are essential to control and manage overall plant performance.

Since knowledge is not only utilized but also continuously gained, shared, updated, and preserved during the operation phase, the KM strategy and the performance of any KM programme need to include all the elements of the knowledge lifecycle that concern all three components of knowledge management.

As an NPP approaches the end of its operating period, the owner/operating organization needs to decide on either ceasing operation or extending the licence period. Consequentially, the extended operation period and/or operation while transitioning to decommissioning are distinctive within the plant operation phase. Therefore, for the purpose of this publication, it is prudent to discuss them separately as long-term operation (i.e. plant operation beyond the original licenced period with extension of licence) and operation while transitioning to decommissioning (i.e. plant operation while preparing for plant disposal).

Annex II discusses specific KM and KM programme performance areas, standards of performance/criteria and Annex V provides examples of KM KPIs during the various stages of plant utilization.

5.3 FACILITY DECOMMISSIONING KNOWLEDGE MANAGEMENT KEY PERFORMANCE INDICATORS

The planning and implementation of decommissioning and waste storage depends significantly on the operating stage and is more easily implemented and managed if all significant and essential information is available. Plant decommissioning usually includes the following stages: preparation for decommissioning, licensing, dismantling, disposal and waste management. IAEA publication NW-T-2.1 [25] describes in detail the development of PIs for decommissioning of nuclear facilities.

During the implementation of decommissioning plans and projects, information and KM affects the successful achievement of the decommissioning strategic objectives which focus on the removal of activated and contaminated materials from the facility, safely and efficiently to:

— Constrain the arisings of radioactive waste to a reasonable minimum;
— Ensure that the decommissioning costs are as low as possible;
— Optimize the clearance or preservation of structures for further unrestricted re-use;
— Protect workers, the public, and the environment from unnecessary radiological exposure.

It is important that decommissioning plans take into account both the most recent and historical knowledge.

Specific training areas to transfer knowledge are:

— Use of a mock-up of a large reactor building to select and provide training on the safest dismantling sequence;
— Simulation of the transfer of activated materials, as part of an ALARA training programme;
— Training on the basic aspects of radiological protection and nationally recognized standards;
— Teaching nuclear and decommissioning fundamentals to young professionals.

From the above lists it is clear that decommissioning KPIs are similar to the KPIs related to the plant operation stage. It is important to note that the decommissioning stage has a smaller number of KPIs but covers a different duration depending on the selected strategy. The critical and most specific KPIs are related to the transition from the operational phase into the decommissioning stage and the final end state after decommissioning is completed. Selected examples of KPIs are provided in of the Appendix .

6. CONCLUDING REMARKS

Implementation of a KM programme will only meet its overall objective of continuously improving the organizational performance, if the improvement and any regression can be easily monitored.

The selection of PIs and the method of reporting them, with regard to the coverage of strategical objectives, will enhance the acceptance of the role of KM in the overall performance of the organization.

A clear, effective, and visual presentation of PIs related to a KM programme will increase their understanding and value for all relevant stakeholders involved in the programme implementation.

The Appendix provides examples of KPIs related to NPPs and organizations with implemented KM programmes but is also relevant as a source of information for other nuclear organizations considering the establishment of a KM programme.

APPENDIX

A.1. LIST OF KEY PERFORMANCE INDICATORS - INTRODUCTION

Key Performance Indicators measure the performance status and progress of programmes, processes, and practices towards the accomplishment of business purposes and the achievement of strategic goals and objectives. The KPIs listed in this appendix are applicable to specific programmes, processes, and practices at different organizational levels of the different phases of the programme implementation and execution in most types of nuclear organizations (NPP operating organizations, nuclear regulators, nuclear technical support/design/consultancy organizations, nuclear R&D organizations, nuclear decommissioning management organizations, nuclear waste management organizations, nuclear education providers, and other nuclear organizations, agencies, or bodies).

The KPIs were developed using the IAEA's KM assessment methodology and KM strategy guide [1]. They are divided into eight main KM functional areas. Each functional area has two categories of KPIs, common KPIs that apply to all phases and specific KPIs that apply only to a specific phase.

A.2. NUMBERING SYSTEM OF KEY PERFORMANCE INDICATORS

All KPIs are numbered for easier referencing, see Fig. A-1. Each number starts with the functional area (i.e., from 1 to 8.), followed by the number indicating the phase (starting with the common KPIs applicable to all phases), e.g. for Functional Area 1:

1. Functional Area:

- 1.1. Area Common KPIs
- 1.2. Project Preparation Specific KPIs
- 1.3. Project Implementation/Plant Realization Specific KPIs
  - 1.3.1. Design. Engineering
  - 1.3.2. Construction
  - 1.3.3. Operation
- 1.4. Decommissioning Specific KPIs
  - 1.4.1. Transition
  - 1.4.2. Decommissioning

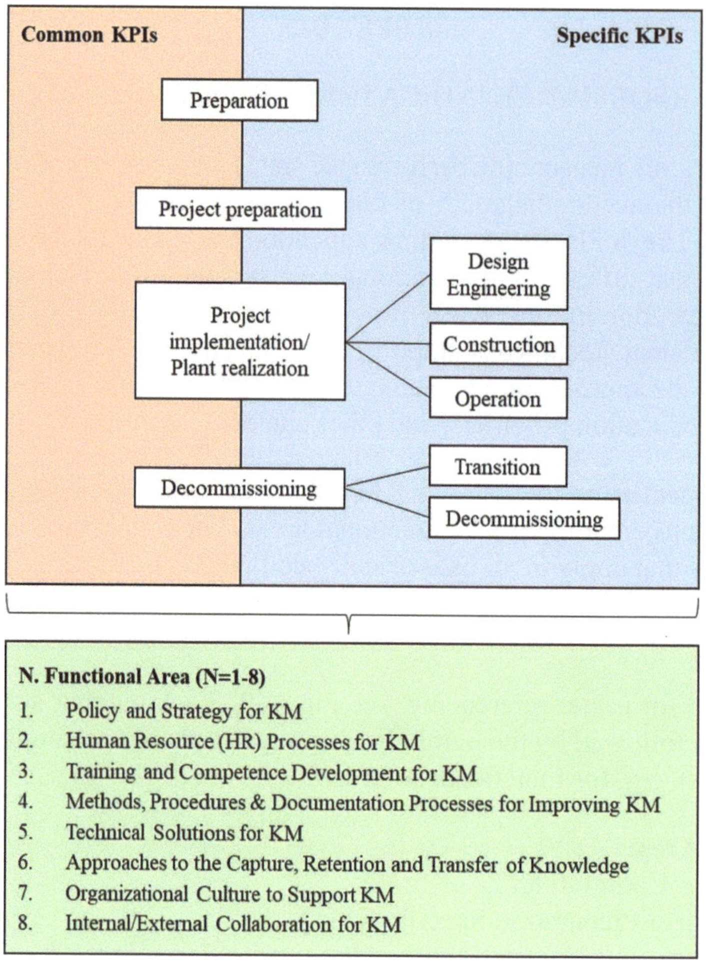

FIG. A-1. Numbering System of the KPIs.

Knowledge Management Indicators Areas

1. **Policy and Strategy Area**
    1.1. Policy and Strategy Area Common KPIs
        1.1.1. NKM policy and strategy exists
        1.1.2. Awareness of the KM policy and its content by managers/staff
        1.1.3. KM policy and strategy communicated to all with clear responsibilities assigned
        1.1.4. NKM strategy addresses the importance of critical knowledge
        1.1.5. Short- and long-term KM goals/plans are developed
        1.1.6. The policy commits the financial and human resource allocation
        1.1.7. KM strategy is incorporated in plant programmes, processes, and procedures
        1.1.8. Percentage of managers/staff who express satisfaction with the implementation of the KM policy/strategy
        1.1.9. Frequency of risk assessment
        1.1.10. Frequency of gap analysis
        1.1.11. Timely and regular assessment of KM programme
    1.2. Project Preparation Specific KPIs
    1.3. Project Implementation/Plant Realization Specific KPIs
        1.3.1. Design and engineering

1.3.2. Construction
1.3.3. Operation
1.3.3.1. Number of KM related non-compliance gaps identified from plant internal audits, reviews, and inspections, as well as reviews from regulatory bodies
1.4. Decommissioning Specific KPIs
1.4.1. Transition
1.4.2. Decommissioning

2. Human Resource Processes for KM
2.1. Human Resource Processes for KM Common KPIs
2.1.1. HRD plan is in compliance with the KM strategy and needs, and is in place and revised periodically
2.1.2. Number of candidate pools to successfully recruit for key positions
2.1.3. Percentage of relevant/involved departments which have implemented KM HRD in their programmes and processes
2.1.4. Percentage of filled expert positions that were identified in the last gap analysis
2.1.5. Percentage of key expert vacancies filled within the expected recruitment time
2.1.6. Turnover rate of in-house and transferred knowledgeable staff
2.1.7. Newly recruited staff turnover rate over a given period (define period)
2.1.8. Number of critical knowledge holders
2.1.9. Percentage of critical knowledge holders at risk
2.1.10. Number of successors identified to receive the knowledge and skill from critical knowledge holders
2.1.11. Job refusal ratio (number of candidates who accepted a job offer/number of candidates who were offered the job)
2.1.12. Number of experienced university lecturers in relevant subjects
2.1.13. Number of entry level nuclear related vacancies and internships in industry, including in academia
2.1.14. Number of professional level nuclear related vacancies in industry, including academia
2.1.15. Percentage of funding for nuclear education and training programmes in the overall budget
2.1.16. Appointment of a highly qualified manager for NKM
2.1.17. Number of successors who succeeded in new positions compared to the number of succession positions where any personal change was implemented within the calendar year (percentage)
2.1.18. Percentage of roles and responsibilities defined
2.2. Project Preparation Specific KPIs
2.2.1. Graduation rate in line with the nuclear infrastructure needs
2.2.2. Percentage of new curricula developed and implemented in line with the infrastructure needs
2.3. Project Implementation/Plant Realization Specific KPIs
2.3.1. Design and engineering

2.3.1.1. Number of recruited experts who have experience and knowledge in project implementation

2.3.2. Construction

2.3.2.1. Number of in-house experts participating in construction phase activities

2.3.2.2. Number of in-house staff trained by external design organizations on construction oversight activities

2.3.2.3. Percentage of construction activities [specified as necessary] performed by in-house staff

2.3.2.4. Number of in-house staff who meet the qualification requirements and are certified in the construction activities phase (specified as necessary)

2.3.2.5. Number of in-house staff in inspection teams for specific area (e.g. a specific SSC) construction, installation, or assembly

2.3.3. Operation

2.3.3.1. Number of vacant SME positions

2.3.3.2. Number of key positions at risk of knowledge loss

2.3.3.3. Number of key positions without identified successors

2.3.3.4. Percentage of key positions with a 'ready now' successor

2.3.3.5. Number of successors employed in a given period of time

2.3.3.6. Retention rate of successors

2.3.3.7. Percentage of role profiles that include awareness of KM as a key requirement

2.4. Decommissioning Specific KPIs

2.4.1. Transition

2.4.1.1. Number of key knowledge holders in total

2.4.1.2. Number of newly nominated key knowledge holders

2.4.1.3. Number of newly nominated key knowledge holders with a high risk of loss

2.4.1.4. Ratio of available and needed internal experts for area X (for example, radiation shielding design basis, ALARA, waste management/storage, HVAC, etc.)

2.4.1.5. Percentage of internal experts needed for decommissioning who are committed (or signed contract) to stay

2.4.2. Decommissioning

2.4.2.1. Number of key decommissioning knowledge holders with a high risk of loss

3. Training and Competence Development for KM

3.1. Training and Competence Development for KM Common KPIs

3.1.1. Percentage of workers who have fulfilled training requirements outlined in their job profiles

3.1.2. Training programme implemented in accordance with organizational objectives to solve identified gaps

3.1.3. Percentage of training sessions delivered by original knowledge holder

3.1.4. Systematic approach to training followed (YES/NO)

3.1.5. Number of staff who attended the training

3.1.6. Organizational learning being incorporated into the training programmes

3.1.7. Percentage of qualified workforce in line with business plan training being met; training documentation being stored; organizational learning being incorporated into the training programmes

3.1.8. Number of exit interviews for knowledge transfer with the critical knowledge holders

3.1.9. Number of critical knowledge holders

3.1.9.1. Number of critical knowledge holders at risk

3.1.9.2. Number of critical knowledge holders' successors prepared

3.1.9.3. Number of experience reports used in training compared to the total number of experience reports

3.2. Project Preparation Specific KPIs

3.3. Project Implementation/Plant Realization Specific KPIs

3.3.1. Design and engineering

3.3.2. Construction

3.3.2.1. Percentage of construction and commissioning documentation transferred from the supplier

3.3.2.2. Number of recruited external experts who worked on the design

3.3.2.3. Percentage of in-house experts joining the team conducting commissioning tests of safety related System, Structure, Components (SSCs) (or specified as necessary)

3.3.2.4. Percentage of in-house experts formally overseeing the commissioning tests of safety related SSCs [or specified as necessary]

3.3.2.5. Number of in-house staff trained by external commissioning organizations on commissioning test of safety related SSCs

3.3.2.6. Percentage of commissioning activities [specified as necessary] performed by in-house staff

3.3.2.7. Percentage of in-house staff who meet qualification requirements and are certified in system operation and testing activities

3.3.2.8. Percentage of test information and knowledge sheets prepared by the in-house staff

3.3.3. Operation

3.3.3.1. Number of knowledge transfer activities undertaken vs planned

3.3.3.2. Number of knowledge retentions performed vs. planned retentions

3.3.3.3. Number of missing or inaccurate data from SSC operation

3.3.3.4. Number of missing or inaccurate data from SSC maintenance and repair

3.3.3.5. Retention rate of the successors

3.3.3.6. Number of successors employed; (period: January-December)

3.3.3.7. Percentage of positions in the successor pool

3.3.3.8. Percentage of designated successor work positions

3.3.3.9. Percentage of non-designated position in the pool

3.3.3.10. Percentage of personnel changed without application

3.3.3.11. Percentage of positions in the succession of which the personnel change took place

3.4. Decommissioning Specific KPIs

3.4.1. Transition

3.4.1.1. Number of internal experts assigned/embedded in the external decommissioning organization(s);

3.4.1.2. Percentage of new staff hired for decommissioning shadowing (or assigned to a mentor-mentee team) internal experts with critical knowledge and confirmed to be departing;

3.4.1.3. Number of training modules in which the critical facility knowledge necessary for decommissioning is covered

3.4.2. Decommissioning

4. Methods, Procedures and Documentation Processes for Improving KM

4.1. Methods, Procedures and Documentation Processes for Improving KM Common KPIs

4.1.1. Number of documents on knowledge-loss risk assessment and mitigation for suppliers, technical support organizations (TSOs), and external service providers

4.1.2. Number of documents on work processes, methodologies and procedures for transferring knowledge, information, and data to/from the vendor, critical equipment/component suppliers, outsourced services, and TSOs

4.1.3. Number of documents on embedded KM principles and practices in its process for collecting and using experience feedback

4.1.4. Percentage of missing or outdated critical knowledge documents

4.1.5. Number of knowledge documents produced from knowledge captured from critical knowledge holders

4.2. Project Preparation Specific KPIs

4.2.1. Licensing and design requirements are well defined, documented, controlled, and retrievable (Yes/No)

4.3. Project Implementation/Plant Realization Specific KPIs

4.3.1. Design and engineering

4.3.1.1. Percentage of design output documents, which are necessary for review and approval by the regulatory body for the construction (or operation) license, turned over from the responsible designer to the project owner

4.3.1.2. Percentage of design documentation containing the basis, criteria, requirements, parameters, codes, and standards for non-safety and plant performance SSCs handed over from the responsible designer to the project owner

4.3.1.3. Percentage of design documentation containing the basis, criteria, requirements, parameters, codes, and standards for non-safety and plant performance SSCs entered into the owner's project information database

4.3.1.4. Number of lost/erased design information records before being entered in the owner's project information database

4.3.1.5. Percentage of recorded documentation noting reason, rationale, and assumptions of why and how design (field) changes have been made (or indicating that there have been no changes)

4.3.1.6. Number of missing or outdated data design documents

4.3.1.7. Percentage of design calculations, drawings, analyses, procurement specifications, and other design documents readily retrievable and clearly describing the bases for the function of plant systems and components

4.3.1.8. Percentage of procedures, drawings, training lesson plans, and related documentation updated promptly following implementation of configuration changes

4.3.1.9. Number of in-house experts participating in vendor design development activities

4.3.1.10. Percentage of design output (specified as necessary) prepared by in-house staff as (on the job training OJT) or on the job participation (OJP)

4.3.1.11. Number of in-house staff trained by external design organizations on design development and implementation activities

4.3.1.12. Number of staff using relevant computer codes in design phase to familiarize themselves with the future state

4.3.2. Construction

4.3.2.1. Percentage of commissioning history dockets and record files transferred and recorded

4.3.2.2. Number of missing or outdated data construction documents

4.3.3. Operation

4.3.3.1. Number of documents for managing records, reports, and dates related to maintenance, surveillance, and inspections

4.3.3.2. Number of records of traceability, rationale, and assumptions of why and how operational, maintenance, and design changes (corporate memory) have been made

4.3.3.3. Percentage of key (Emergency Response Organization (ERO) positions covered by at least one successor (rule and metrics: ≥ 90%)

4.3.3.4. Training schedule adherence (percentage of training session completed as scheduled)

4.3.3.5. Training readiness (number of missed training preparation milestones for scheduled training)

4.3.3.6. Number of issues (near misses, etc.) related to knowledge deficiency

4.3.3.7. Number of exams not passed due to lack of knowledge

4.3.3.8. Number of coaching/mentoring programmes initiated

4.3.3.9. Coaching/mentoring ratio of the management (number of mentoring/coaching hours per management level over a defined period)

4.3.3.10. Number of missing or outdated data operating documents

4.3.3.11. Plant procedures, particularly emergency and abnormal procedures, are written in accordance with applicable owner’s group guidance, plant-specific guidelines, plants specific probabilistic safety analyses, and vendor technical manuals (Yes/No)

4.4. Decommissioning Specific KPIs

4.4.1. Transition

4.4.1.1. Number of internal experts assigned/embedded in external decommissioning organization(s);

4.4.1.2. Percentage of new staff hired for decommissioning shadowing of internal experts with critical knowledge and confirmed to be departing

4.4.1.3. Number of training modules in which the critical facility knowledge necessary for decommissioning is covered

4.4.2. Decommissioning

4.4.2.1. Number of missing or outdated decommissioning documents

5. Technical Solutions for KM

5.1. Technical Solutions for KM Common KPIs

5.1.1. Percentage of documents received and recorded in the configuration database

5.1.2. Percentage of programme/project/plant documents recorded and stored in accordance with organizational business objectives

5.1.3. Number of staff using relevant computer codes to familiarize themselves with the future state

5.1.4. Number of inhouse experts on a specific software/code that is critical for current and future phases

5.1.5. Number of documents uploaded to the knowledge portal

5.1.6. Number of uploaded experience reports to the knowledge portal

5.1.7. Number of visitors accessing the knowledge portal, library, etc.

5.2. Project Preparation Specific KPIs

5.3. Project Implementation/Plant Realization Specific KPIs

5.3.1. Design and engineering

5.3.1.1. Percentage of computer codes transferred by the designer and verified by the in-house staff

5.3.1.2. Number of code manuals and standard test cases

5.3.2. Construction

5.3.3. Operation

5.3.3.1. Number of IT support tools' users

5.3.3.2. Usage time of IT support tools per user

5.3.3.3. Unscheduled downtime of an IT support tool

5.3.3.4. Number of attempts to find relevant information

5.3.3.5. Percentage of emergency drills in which new Emergency Response Team (ERT) member for ERT position X shadowed the expert ERT member

5.4. Decommissioning Specific KPIs

5.4.1. Transition

5.4.2. Decommissioning

6. Approaches to the Capture, Retention and Transfer of Knowledge

6.1. Approaches to the capture, retention, and transfer of knowledge common KPIs

6.1.1. Number of local experts participating in quarterly meetings of the external report provider organization
6.1.2. Number of local experts embedded in the external report provider organization
6.1.3. Number of errors or deficiencies identified by local experts in the work of international experts
6.1.4. Number of training session by international experts for the local experts
6.1.5. Number of local experts participating in joint meetings
6.1.6. Number of local experts embedded in the external report provider
6.1.7. Number of local experts trained by international experts
6.1.8. Number of reviews undertaken by local experts where area for improvements were identified
6.1.9. Number of training sessions by the international experts for the local experts
6.1.10. Percentage of relevant/involved organizations which have implemented the KM policy and strategy
6.1.11. Percentage of design output documents, which are necessary for review and approval by the regulatory body for construction (or operation) license, handed over from the responsible designer to the project owner
6.1.12. Number of in-house experts participating in off-site vendor design development activities
6.1.13. Percentage of design output (specified as necessary) prepared by in-house staff as OJT or OJP
6.1.14. Number of in-house staff trained by external design organizations on design development and implementation activities
6.1.15. Percentage of design documentation containing the basis, criteria, requirements, parameters, codes and standards for non-safety and plant performance SSCs turned over from the responsible designer to the project owner
6.1.16. Percentage of design documentation containing the basis, criteria, requirements, parameters, codes, and standards for non-safety and plant performance SSCs entered in the owner's project information database
6.1.17. Number of lost/erased design information records before being entered in the owner's project information database
6.1.18. Percentage of recorded documentation noting reason, rationale and assumptions of why and how design (field) changes have been made (or indicating that there have been no changes)
6.1.19. Percentage of realized knowledge transferred (any KM tool used) to the number of knowledge employees who left during the calendar year

6.2. Project Preparation Specific KPIs

6.3. Project Implementation/Plant Realization Specific KPIs

6.3.1. Design, and engineering
6.3.2. Construction
6.3.3. Operation

6.4. Decommissioning Specific KPI

6.4.1. Transition

6.4.1.1. Number of Communities of Practice (CoP)

6.4.1.2. Number of expert CoP KPIs related to succession

6.4.1.3. Number of experience reports used in training compared to the total number of reports

6.4.1.4. Number of experience reports

6.4.1.5. Number of realized knowledge transfers (any KM tool used)

6.4.1.6. Number of knowledgeable employees who left during the calendar year

6.4.1.7. Successors′ score

6.4.1.8. Percentage of documents collected and archived during the operation that are identified as decommissioning related;

6.4.1.9. Percentage of documents collected and archived during post-shutdown that are identified as decommissioning related;

6.4.1.10. Number of records of legacy radioactive waste inventory materials and their locations;

6.4.1.11. Percentage of dedicated staff (identified as critical to the decommissioning programme, such as chemical, radiation protection, experts 'activities, etc.) participating in training sessions

6.4.2. Decommissioning

7. Organizational Culture to Support KM

7.1. Organizational Culture to support KM Common KPIs

7.1.1. Number of organized events for newcomers (mentoring programme)

7.1.2. Percentage of managers encouraging cooperation and teamwork among plant organizational units, especially when the successful implementation of work activities requires support from several groups

7.1.3. The organization's values and behaviours for transfer of knowledge are modelled by its leaders and practiced by all plant staff (Yes/No)

7.1.4. Effective mechanisms are in place to promptly transfer the organization's values and expectations regarding transfer of knowledge to new staff (Yes/No)

7.1.5. Managers practice visible leadership by personally observing performance, coaching, mentoring, and reinforcing standards (Yes/No)

7.2. Project Preparation Specific KPIs

7.3. Project Implementation/Plant Realization Specific KPIs

7.3.1. Design and engineering

7.3.2. Construction

7.3.3. Operation

7.4. Decommissioning Specific KPIs

7.4.1. Transition

7.4.2. Decommissioning

8. Internal and External Collaboration for KM

8.1. Internal and External Collaboration for KM Common KPIs

8.1.1. (Percentage) ratio of participants from other organizations to participants in the managing (core) entity (openness of CoP)

8.1.2. Number of knowledge assets produced (depending on the nature of the CoP and its purpose, the number of artefacts, e.g. drawings, documents, products etc. the CoP has produced that have been applied to specific business needs)

8.1.3. Number of case studies produced by each CoP describing the benefits of its work

8.1.4. Percentage of CoP satisfaction (may be part of a general employee satisfaction survey)

8.2. Project Preparation Specific KPIs

8.3. Project Implementation/Plant Realization Specific KPIs

8.3.1. Design and engineering

8.3.2. Construction

8.3.3. Operation

8.3.3.1. Number of active CoPs in place

8.3.3.2. Number of success stories of CoPs by year (sharing useful knowledge)

8.3.3.3. Number of after-action reviews (number of follow-up reviews)

8.3.3.4. Number of external reports shared within the organization

8.3.3.5. Number of employees participating in workshops, conferences, peer reviews, and missions

8.3.3.6. Number of benchmarking initiatives performed over a period of time

8.3.3.7. Number of analyzed and shared operating events received from external sources

8.3.3.8. Number of corrective measures initiated based on event analysis

8.3.3.9. Number of low-level events and near misses related to human performance and organizational factors in possessing or using critical knowledge

8.3.3.10. Percentage of analyzed events caused by human error and inadequate procedure in using critical knowledge

8.3.3.11. Number of situations that caused delay/cancellation/failure/errors in jobs or activities due to lack of critical knowledge holders

8.3.3.12. Number of staff engaging in CoP activities (simple measure of total numbers or percentage of staff signed-up to and actively involved in the activities of the CoP)

8.4. Decommissioning Specific KPIs

8.4.1. Transition

8.4.1.1. Number of CoP closures (much later in the programme some CoPs will reach the end of their useful lifetime and will cease to add value. It's important that these do not act as a drain on staff resource, money and time).

8.4.1.2. Decommissioning

REFERENCES

[1] WORLD ASSOCIATION OF NUCLEAR OPERATORS, Performance Objectives and Criteria, WANO PO&C 2013-1, WANO, London (2013).

[2] INTERNATIONAL ATOMIC ENERGY AGENCY, Guide to Knowledge Management Strategies and Approaches in Nuclear Organizations, IAEA Nuclear Energy Series, No. NG-G-6.1, IAEA, Vienna (2022).

[3] INTERNATIONAL ATOMIC ENERGY AGENCY, Development and Implementation of a Process Based Management System, IAEA Nuclear Energy Series, No. NG-T-1.3, IAEA, Vienna (2015).

[4] INTERNATIONAL ATOMIC ENERGY AGENCY, Leadership and Management for Safety, IAEA Safety Standards Series No. GSR Part 2, IAEA, Vienna (2016).

[5] INTERNATIONAL ATOMIC ENERGY AGENCY, The Management System for Nuclear Installations, IAEA Safety Standards Series No. GS-G-3.5, IAEA, Vienna (2009).

[6] INTERNATIONAL ATOMIC ENERGY AGENCY, Application of the Management System for Facilities and Activities, IAEA Safety Standards Series No. GS-G-3.1, IAEA, Vienna (2006).

[7] INTERNATIONAL ORGANIZATION FOR STANDARDIZATION, Knowledge Management Systems – Requirements, ISO Standard No. 30401, ISO, Geneva (2018).

[8] INTERNATIONAL ATOMIC ENERGY AGENCY, Knowledge Management and Its Implementation in Nuclear Organizations, Nuclear Energy Series, No. NG-T-6.10, IAEA, Vienna (2016).

[9] PARMENTER, D., Key Performance Indicators: Developing, Implementing and Using Winning KPIs, 3rd Edition, John Wiley & Sons, Inc., Hoboken, New Jersey, (2015) 307 pp.

[10] INTERNATIONAL NUCLEAR SAFETY ADVISORY GROUP, Management of Operational Safety in Nuclear Power Plants, INSAG-13, IAEA, Vienna (1999).

[11] DORAN, G. T., “There's a S.M.A.R.T. way to write management’s goals and objectives”. Management Review. 70 (11): 35–36 (1981).

[12] INTERNATIONAL ATOMIC ENERGY AGENCY, Operational Safety Performance Indicators for Nuclear Power Plants, IAEA TECDOC Series TECDOC-1141, IAEA, Vienna (2000).

[13] INTERNATIONAL ATOMIC ENERGY AGENCY, Economic Performance Indicators for Nuclear Power Plants, IAEA Technical Reports Series No. 437, IAEA, Vienna (2006).

[14] SPITZER, D. R., Transforming Performance Measurement: Rethinking the Way We Measure and Drive Organisational Success, HarperCollins Focus, Nashville (2007).

[15] INTERNATIONAL ATOMIC ENERGY AGENCY, Planning and Execution of Knowledge Management Assist Visits for Nuclear Organisations, IAEA-TECDOC-1880, IAEA, Vienna (2019).

[16] INTERNATIONAL ATOMIC ENERGY AGENCY, Knowledge Management for Nuclear Research and Development Organizations, IAEA TECDOC Series TECDOC-1675, IAEA, Vienna (2012).

[17] INTERNATIONAL ATOMIC ENERGY AGENCY, Technical Support to Nuclear Power Plants and Programmes, IAEA Nuclear Energy Series No. NP-T-3.28, IAEA, Vienna (2018).

[18] INTERNATIONAL ATOMIC ENERGY AGENCY, Project Management in Nuclear Power Plant Construction: Guidelines and Experience, IAEA Nuclear Energy Series No. NP-T-2.7, IAEA, Vienna (2012).

[19] INTERNATIONAL ATOMIC ENERGY AGENCY, Milestones in the Development of a National Infrastructure for Nuclear Power, IAEA Nuclear Energy Series No. NG-G-3.1 (Rev. 1), IAEA, Vienna (2015).

[20] INTERNATIONAL ATOMIC ENERGY AGENCY, Responsibilities and Capabilities of a Nuclear Energy Programme Implementing Organization, IAEA Nuclear Energy Series No. NG-T-3.6 (Rev. 1), IAEA, Vienna (2019).

[21] INTERNATIONAL ATOMIC ENERGY AGENCY, Fundamental Safety Principles, IAEA Safety Standards Series No. SF-1, IAEA, Vienna (2015).

[22] INTERNATIONAL ATOMIC ENERGY AGENCY, Power Reactor Information System (PRIS), https://www.iaea.org/pris/

[23] INTERNATIONAL ATOMIC ENERGY AGENCY, Plant Life Management Models for Long Term Operation of Nuclear Power Plants, IAEA Nuclear Energy Series No. NP-T-3.18, IAEA, Vienna (2015).

[24] INTERNATIONAL ATOMIC ENERGY AGENCY, SALTO Peer Review Guidelines, IAEA Services Series No. 26, IAEA, Vienna (2013).

[25] INTERNATIONAL ATOMIC ENERGY AGENCY, Decommissioning of Nuclear Facilities: Training and Human Resource Considerations, IAEA Nuclear Energy Series No. NG-T-2.3, IAEA, Vienna (2008).

[26] INTERNATIONAL ATOMIC ENERGY AGENCY, Selection and Use of Performance Indicators in Decommissioning, IAEA Nuclear Energy Series, No. NW-T-2.1, Vienna (2011).

[27] INTERNATIONAL ATOMIC ENERGY AGENCY, Decommissioning of Facilities, IAEA Safety Standards Series No. GSR Part 6, IAEA, Vienna (2014).

[28] OZ MINERALS, Global Performance Standards, OZ Minerals, Melbourne (2021)

[29] R. Johnston, S. Brignall and L. Fitzgerald. 'Good Enough' Performance Measurement: A Trade-Off between Activity and Action. The Journal of the Operational Research Society. Vol. 53, No. 3, Part Special Issue: Performance Management (2002), pp. 256-262

[30] ISO 30401:2018(en) Knowledge Management Systems — Requirements

[31] ISO 9000:2015, Quality management systems — Fundamentals and vocabulary

ANNEX I.

PERFORMANCE INDICATORS WITHIN THE MANAGEMENT SYSTEM

The strategic goals of a nuclear organization are established to enable the organization to achieve its purpose and in doing so, address all its requirements and expectations, including those regarding safety, reliability, quality, security, safeguards, environment, and health as well as those with legal, financial, and economic elements and the organizational culture for safety and performance.

During an organization's lifetime, numerous strategic decisions are made to achieve its purpose effectively and efficiently while meeting its requirements and expectations. To make good decisions in the organization, we need a system that considers all important factors together, instead of making decisions in isolation. An integrated management system (IMS) provides this single framework to manage the arrangements, programmes, processes, and resources necessary to address all the decision-making goals of the organization.

The IMS also has to ensure the regular monitoring, measuring, and assessment of KM practices and programmes with the aim of overall improvement, as well as improvements of human performance by identifying the areas requiring improvement (e.g. correction, strengthening, support) and by determining the means to achieve and confirm the effectiveness of such improvements. Accordingly, the IMS supports a proactive and responsive management of the arrangements, programmes, processes, and resources, which includes the knowledge and the information of the organization[I-1].

In an effective management system, the KM PIs are therefore directly linked to the organization's strategic objectives as inputs to its decisions making process. The status and trend of KM performance (and its integrated impact on the performance of other programmes, processes, and procedures) enables sound decision based on overall performance. Besides the establishment of a clear link to the strategic direction of the organization, it is also assumed that the organization follows a structured 'plan, do, check, act, improve' (PDCA) [I-2] cycle as shown in Fig. I-1.

The PDCA cycle is an industry recognized continuous improvement tool that provides a four-step method for the control and improvement of a process and a product, which can easily be applied to measuring and improving the effectiveness of a KM programme and associated processes and practices.

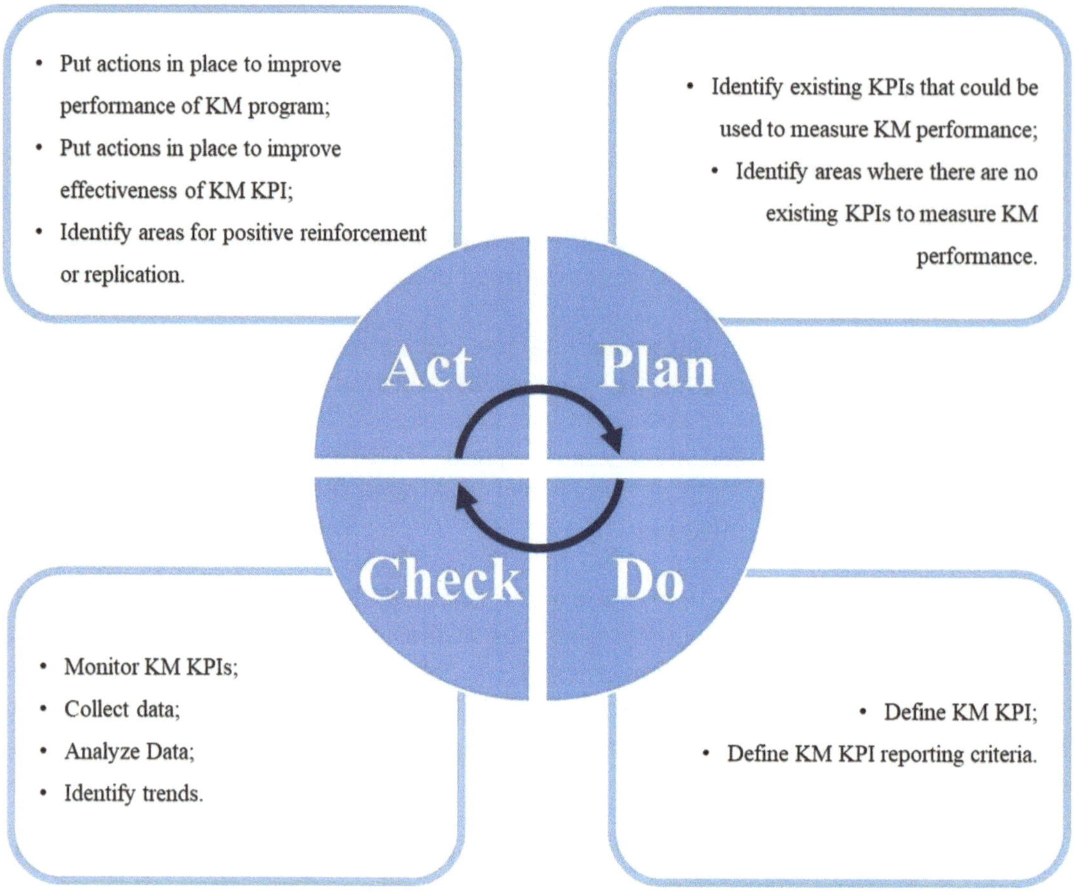

FIG. I-1 KM Plan, Do Check, Act (PDCA) Cycle.

I-1. PLAN

It is important to ensure that the KPIs are aligned with the main organizational objectives. Find and use existing Key Performance Indicators (KPIs) to measure a Knowledge Management (KM) Programme. If no suitable KPIs exist, identify performance areas and create KM KPIs as needed.

Example of an existing KPI: Completion rates for staff appraisals are a common KPI and can be utilized in a KM context. along with setting an expectation that KM is discussed during appraisals, the extent of the KM discussions held during the appraisals can also be measured as a KPI.

Example of a required KM KPI: A key element of KM is understanding where the critical knowledge is being held. A KM KPI may need to be developed which measures the number of identified critical knowledge loss risk areas.

I-2. DO

When defining new KM KPIs, data and information relating to the chosen KPIs need to be:

— From a reliable source and, where possible, readily available in existing information systems such as enterprise resource planning (ERP), financial, human resource or other similar systems;
— Regular and time dependent – trending data with time is useful as it shows how the KPI is progressing and whether real progress is being made;
— Simple and representing the measurement needed to demonstrate success. Parameters derived from various data sets are often subject to scrutiny and are less likely to be believed;

— Automated where possible, ensuring the quality of data accuracy and minimizing additional workload.

After defining a KM KPI the best method of measuring and assessing performance against it needs to be determined. In some cases, it may be more effective to break down the KPI into more manageable components and measure each part separately.

Using the example above of understanding where critical knowledge is held, the KPI would be more effective if split into two components:

— Completion rate of KM risk analysis;
— Number of critical knowledge risk areas.

This allows each element to be measured and addressed separately as, although linked to overall performance area, each element would require a different approach if an improvement need was identified.

I-3. CHECK

For KPIs to work as a management tool and ensure the effective oversight and continuous improvement of KM programmes, they have to be accurately measured and reported. This activity can be split into three main phases:

— Monitor KM KPIs;
— Analyse data;
— Identify trends.

It is also important to monitor the effectiveness of the KM KPIs as areas being measured within a KM programme may be behavioural rather than data led and, as such, difficult to define.

Example of a behavioural area: Following the expectation that KM is discussed at staff annual appraisals, the existing annual appraisal completion rate KPI may be used to monitor the performance in this area. Whilst the completion rate KPI will indicate the number of appraisals completed, it will not necessarily correlate with the number of KM discussions held, as this depends on the behaviour of the individual leaders conducting the appraisal. One simple way to test the effectiveness in this case would be to conduct a short survey asking appraised staff if KM was discussed at their last annual appraisal.

KPI data need to be presented to senior management in a clear and concise manner emphasizing the benefits of the KM programme or initiative. This aspect needs to be a regular agenda item when discussing the initiative with senior management as part of a monthly review.

I-4. ACT

For the PDCA cycle to be effective within a KM programme, it is imperative that the areas identified for improvement are acted upon and then measured for effectiveness during the next reporting cycle.

Knowledge Management action plans need to be developed and rigorously implemented to:

— Improve performance of the KM programme;
— Improve effectiveness of the KPIs.

It is also important to identify areas within a KM programme that have strong effect on positive reinforcement and replication. This, along with acting on areas for improvement, forms the basis of a strong continuous improvement process.

I-5. PERFORMANCE INDICATORS WITHIN THE BUSINESS CASE

KM focuses on leveraging an organization's knowledge assets to create and sustain value for the business, its employees, and its customers. An effective KM programme that is aligned with the organization's business profile, its strategic goals, and success factors ensures that knowledge that is critical for the performance of the required and expected activities is acquired, preserved, and utilized towards its business purpose and objectives, particularly recognizing that:

- High business performance is assured by well-informed decisions being made at every level of the organization, which requires the use of the necessary knowledge known by high-level/high-quality experts and that the loss of critical knowledge by the departures of high-level/high-quality experts from the organization adversely affects the overall performance;
- Non-existing or insufficient knowledge reduces the effectiveness of strategic and tactical planning and the organization's business focus and that this hinders the productivity and efficiency of essential activities owing to short or long pauses for rethink, rework or failures;
- Institutional knowledge drives innovation protects intellectual capital and improves the company's methods and tools to achieve its purpose more cost-effectively and efficiently with the best return on investment, particularly those involving the introduction and maintenance of new technologies and organization-wide knowledge requiring the costly change of processes, practices, and behaviour.

As a part of an effective conduct of business, KM PIs are directly linked to the organization's strategic objectives to measure actual performance against the expected performance standards and to provide an input to make informed decisions. The status and trends of performance in managing knowledge (and its integrated impact on the performance of other programmes, processes, and procedures) that is aligned with the organization's business profile, its strategic goals and its critical success factors consequently will:

- Enable sound decision making on the KM programme based on overall organization performance, as KM focuses on leveraging an organization's knowledge assets to create value for the business, its employees and its customers;
- Ensure the efficient accomplishments of business specific functions which are related to/impacted by knowledge aspects by clearly recognizing the integration or reflection of KM initiatives in the overall corporate strategy.

I-6. PERFORMANCE INDICATORS FOR CONTINUOUS IMPROVEMENT

Management and other staff need to be aware and informed of the strengths and weaknesses of their own organization, other organizations and interfaces in establishing, administering, maintaining and applying KM and KM programmes. A lack of an effective and structured process to identify, analyse, correct, and keep records of issues, deficiencies and weaknesses in KM programmes, processes and practices would result in an accumulation of deficiencies in the conduct of and processes for knowledge management. Consequently, this accumulated deficiencies, some of which would go unnoticed until an impact appears, could lead to significant errors and events resulting from a lack, or unavailability, of critical knowledge that would jeopardize safety and reliability and/or the efficient and effective performance of the organization.

Therefore, the management in the organization needs to proactively check, identify, and correct weaknesses in the KM programmes and associated programmes, processes, practices and procedures, which requires their (and relevant plant staff's) active support for routinely monitoring and assessing (as well as observing and reporting) to identify any areas for improvement and the areas of strength, to improve or to sustain performance, respectively.

More importantly, a continuous improvement process also necessitates the initiation, application and improvement of tangible measures and metrics to assess the status of KM and KM programmes, as to their establishment, implementation and practices regarding the detection of declining or improving performance or measuring effectiveness. In order to do so, PIs are needed to identify whether the activities relating to organizational knowledge are deteriorating or are being maintained or improving.

REFERENCES

[I-1] INTERNATIONAL ATOMIC ENERGY AGENCY, Leadership and Management for Safety, IAEA Safety Standards Series No. GSR Part 2, IAEA, Vienna (2016).

[I-2] ISO 9000:2015, Quality management systems — Fundamentals and vocabulary

ANNEX II.

KNOWLEDGE MANAGEMENT PERFORMANCE AREAS

The suggested framework for KM KPIs has been developed across eight KM functional areas, which are described in the IAEA publication NG-G-6.1. This Annex provides more details on the eight functional areas:

— Policy and strategy;
— Human resources processes for KM;
— Training and competence development for KM;
— Methods, procedures, and documentation processes for improving KM;
— Technical solutions for KM;
— Approaches to capturing, retaining, and transferring knowledge;
— Organizational culture to support KM;
— Internal/external collaboration for KM.

II-1. POLICY & STRATEGY FOR KNOWLEDGE MANAGEMENT

Every KM system needs to have a written policy, communication strategy, implementation strategy, and identified responsibilities and accountabilities.

With the aim of improving organizational performance, it is imperative that the strategy and policy for KM is part of a wider organizational strategy and does not stand alone, being clearly aligned with the strategic objectives, supporting existing policies (such as safety culture policy), with ownership at a senior level.

To support effective integrated knowledge management, knowledge needs to be recognized as an asset by the whole organization with employees understanding their roles and responsibilities in achieving this goal. To fully realize this, a diverse and inclusive culture needs to be reinforced, along with the allocation of both adequate human and financial resources.

It is important to consider this area annually when the financial budgets are set, to ensure that KM is fully integrated to support the business vision and strategic objectives and that it maintains alignment as the business moves through different stages – making the business both effective and sustainable.

II-2. HUMAN RESOURCES PROCESSES FOR KNOWLEDGE MANAGEMENT

To meet the strategic objectives of the organization, KM may be considered in all HR processes throughout the different stages of the employee lifecycle as they advance through an organization (attract, recruit, onboard, develop, retain, and separate).

This is a very important area to ensure that the organization has the right people with the right skills at the right time to keep the organization safe and delivering on its objectives.

II-2.1. Recruit

A strong workforce planning methodology provides an indication of the skills required for the business and allows the right people to be recruited. This needs to be documented in job profiles or equivalent. As the recruitment process can be lengthy, monitoring this ensures that the necessary steps are taken at an early stage, allowing sufficient time to attract and recruit the right people.

II-2.2. Train and Develop

Identifying core technical skills along with the qualifications and competence requirements is the core to successful knowledge management. Having a strong understanding of the people and roles in the organization means appropriate action can be taken to manage knowledge and effective succession planning can be put in place to minimize the risk of knowledge loss. Again, this can be a lengthy process, so it is important to ensure that PIs are in place to highlight any issues and allow for appropriate planning.

II-2.3. Retain

HR practices and processes need to be in place to support retention of employees, with clear ownership.

Motivating employees to stay within the organization and, therefore, retaining their knowledge is an effective way to support knowledge management. Annual appraisals are one way to ensure KM objectives are documented and understood. PIs in this area help retain and improve the value of existing knowledge.

II-2.4. Exit

All employees leave the organization at some point. KM needs to apply adequate interventions with the aim of capturing, storing, and sharing the knowledge as individuals move through and out of the organization. PIs in this area indicate potential risks and required mitigating actions to preserve knowledge, before it leaves the organization.

II-3. TRAINING AND COMPETENCE DEVELOPMENT FOR KNOWLEDGE MANAGEMENT

Ensuring that people are adequately trained and competent to perform their role is crucial for organizational success. Initial and continuous training can be conducted in a systematic way, ensuring that the business objectives are being considered and delivered using appropriate and up-to-date methods and supported by the use of subject matter experts, experienced mentors, and coaches.

This can encourage knowledge sharing between roles but also across generations. In addition, it can provide information and guidance from experts, support the alignment of career paths, and increase employee engagement and performance.

Measuring performance in this area can visibly demonstrate to all stakeholders the importance of sharing knowledge, which is fundamental to KM and consequently to organizational success.

II-4. METHODS, PROCEDURES AND DOCUMENTATION PROCESSES FOR IMPROVING KNOWLEDGE MANAGEMENT

The methods, procedures, and documentation for improving KM in an organization are identified and implemented in order to support strategic organizational objectives.

This can include the methodical identification of the knowledge that is critical for the organization, ensuring that technical drawings, codes, work activities, lesson plans, lessons learned, etc. are kept up-to-date and also conducting self-assessments to understand progress and areas for improvement, as well as internal and external audits to improve organizational effectiveness.

Ensuring that critical historical data, information, and knowledge is retained as the organization moves through different phases, including the storage and retrieval of this information in a well-documented way, supports efficiency, and effectiveness as well as safety.

Performance Indicators provide an insight into the effectiveness and efficiency of these methods and procedures as well as the quality of documentation and processes, allowing for analysis to support continual improvement.

II-5. TECHNICAL SOLUTIONS FOR KNOWLEDGE MANAGEMENT

The application and integration of information strategies, systems, and technologies that support KM such as databases, content and document management systems, use of the internet and social networking technologies is crucial for a successful KM programme.

It is important that KM processes are enabled by IT solutions – for capturing, accessing, and ensuring the integrity of information.

Performance Indicators can be used to measure the effectiveness of IT solutions and IT solutions can be used to facilitate KPI measurement. A typical IT system used to support KM could include:

— KM databases;
— Simulation tools (codes, analytical models, etc.);
— Enterprise resource planning (ERP) systems;
— Portals/intranets;
— Knowledge search engines;
— Expert yellow pages (directory);
— Expert/intellectual systems;
— E-learning platforms;
— Wikis/blogs;
— Plant information modelling/building information modelling.

It is important that the PIs in this area support the reliability and, therefore, the effectiveness of these solutions.

II-6. APPROACHES TO CAPTURE/TRANSFER OF KNOWLEDGE

Having efficient and effective approaches in place to capture and transfer knowledge underpins the success of KM.

These approaches may include on-boarding, succession planning, pre-job briefing, and post-job debriefing, paired working/shadowing, elicitation and exit interviews, root cause analysis, and video capturing to preserve knowledge as well as coaching and mentoring techniques.

Appropriate KM tools utilized to capture and transfer knowledge could include:

— Knowledge and risk mapping;
— Concept sorting and mapping;
— Process mapping;
— Explicit knowledge capture (narrative documentation);
— Story telling.

Measuring PIs in this area, can reassure the organization that knowledge is being effectively captured and transferred, providing the opportunity to improve the approaches if required.

II-7. ORGANIZATIONAL CULTURE TO SUPPORT KNOWLEDGE MANAGEMENT

It is important that organizations have a strong knowledge sharing culture, which supports KM practices. This culture needs to be driven by the senior management of the organization, with leaders minimizing the risk of knowledge loss within their teams in line with organizational objectives.

An open culture needs to be actively encouraged allowing for the transparent reporting of events so lessons may be learned, and appropriate corrective action may be taken supported by an effective programme to track and monitor these actions.

Appropriate PIs measure if the culture is supporting KM in an open and proactive way and allow interventions for improvement as required.

II-8. INTERNAL AND EXTERNAL COLLABORATION FOR KNOWLEDGE MANAGEMENT

Collaboration can flourish if the organizational culture supports collaboration with the relevant internal and external organizations/departments/stakeholders responsible for its strategic knowledge management, e.g. through setting up 'Communities of Practice' (CoPs).

PIs in this area may measure external benchmarking or the efficiency of formal and informal processes in place to capture and transfer knowledge and information from critical suppliers, outsourced service providers or technology vendor countries.

ANNEX III.

EXAMPLES OF LINKING KEY PERFORMANCE INDICATORS TO ORGANIZATIONAL GOALS AND STRATEGIES

This Annex provides several examples to support an overall approach for identifying KM KPIs and KM-related KPIs inside an organization's performance assessment framework as described in the Section 4.1.

III-1. AN EXAMPLE FROM THE PAOLO VERDE NUCLEAR POWER PLANT

The examples shown in Fig. III-1 and in Tables from III-1 to III-3 include four organizational levels (tiers). It starts with organizational goals and objectives, continues with the key priorities to reach these goals and metrics to measure KM performance on different organizational levels.

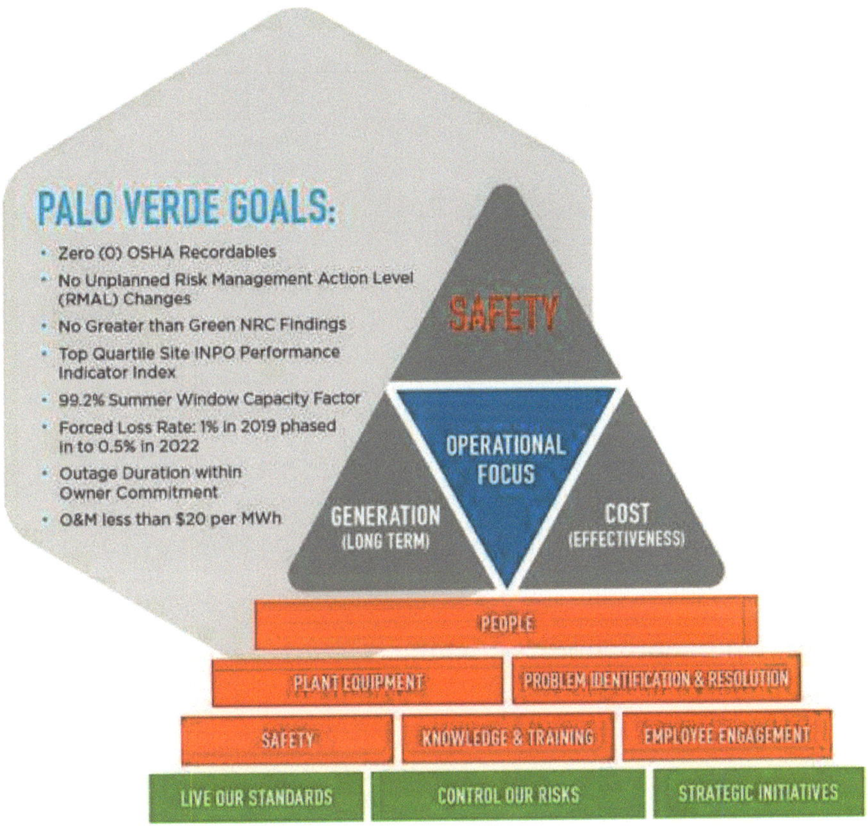

FIG. III-1. A specific example of linking KPIs to organizational goals and objectives.

TABLE III-1. KEY PRIORITIES

Palo Verde Building Blocks	Management Oversight Forum	Business Area Key Priorities
Knowledge & Training	Senior Training Council	Knowledge is the foundation upon which we build. We continue to build our knowledge on a daily basis. We value knowledge, learning and training as key strategic tools to develop our workforce and improve overall station performance. The focus placed on knowledge and training is emphasized by it being one of Palo Verde's three aspirations to lead the industry. Training is used as a strategic tool to provide highly skilled and knowledgeable personal for safe, reliable operations and support performance improvement.

TABLE III-2. METRICS SUMMARY

Palo Verde Building Blocks	People	Plant Equipment	Problem Identification & Resolution	Safety	Knowledge & Training	Employee Engagement	Live Our Standards & Control Our Risks
Tier 3	• APTMS Development Goals • Succession Planning • Pipeline Program Retention rate • Mentoring	• On-Line Reliability Loss Factor • Chemistry Effectiveness Index • Mitigating System Performance Index • CY Metric (Yellow to Green)		• Level 1 and Level 2 Personnel Contamination Events • E-Plan Health • Physical Protection • TISAR • Unplanned Risk Management Action Level Changes • Reportable Environmental Incidents	• Knowledge Transfer and Retention Plans • Training Programme Health • Operations Training Program Accreditation Renewal Readiness • TR AFI (OJT/TPE)	• Nuclear Professional AFI Closure	• Phishing Catch Rate • Access Revocation Time • No Greater Than Green NRC Violations • Operational Risk D-M AFI Closure • Industrial Safety AFI Closure • Site Clock Reset (excluding industrial safety)
Tier 4	• Leader Development Fundamentals • Leader Development Academy		• Condition Adverse to Quality (CAQ) and Significant CAQ >1080 Days Old • NCAQ evaluations > 30 days	• Safety Culture Priority Groups • Hazardous Waste • Non-reportable Environmental Incidents			

TABLE III-3. METRIC TARGETS

Tier	Metric Name	Description	2018 Goal	2018 CWP	2019 Goal	2019 CWP	2020 Target
3	E- Plan Health	Combines results of the STARS EP Lower-Level indicators with NRC EP Cornerstone PI into a single measure of program health.	≥ 90	92.5	≥ 90		
3	Physical Protection	Intrusion Detection Zone and Closed-Circuit TV unavailability index derived from specific compensatory hours. The metric is a rolling 12-month indicator of performance for Physical Protection. PVGS has established performance thresholds that will turn our station metric red at 0.04, which would be halfway to the NRC concern area of 0.08; a performance index greater than 0.039 requires management attention.	≤ .03	.004	≤ .03		
3	Knowledge Transfer and Retention	Ensure Knowledge Transfer and Retention for the site by ensuring the plans are on-track or complete.			≥ 90%		
3	Training Program Health	Trends the aggregate performance of Training programs as indicated by various performance measures indicating program ownership and engagement, program performance, and station performance.	≥ 90	93.4	≥ 90		≥ 90
3	Phishing Catch Rate	Percentage of employees who did not recognize a phishing attempt and clicked a link or opened an attachment in the email	5%		5%		
3	Access Revocation Time	The average number of days taken for leaders to revoke access in response to employee/ contractor separation. Regulatory compliance requires computer and physical access to be revoked within 24 hours of employee/contractor's employment with APS ending.	< 24 Hours		< 24 Hours		

Approach of building organization's PIs and KPIs at different levels is described in the Figure III-1 and comprehensively illustrated in the Table III-3. The single example given below support an overall approach for identifying KM KPIs and KM-related KPIs inside the organization's performance assessment framework. Approach illustrated in the Table III-3 is converted into a practical example in the Figure III-2.

Example includes identification of organization's strategic goals and objectives as well as global and specific performance area related to KM. Within this hierarchical framework of KM performance and associated PIs in the context of level of decision making, KM attributes and PIs needs represent past, current, and future status and performance at a particular level (i.e. of a strategy, policy level, process or activity in a weighted manner).

Strategy:	Safe operation of nuclear plants
Strategic Objective:	Safety – with the attribute, nuclear safety
Global performance area:	Emergency response
Specific performance area:	Emergency exercise – Competency building
Activity performance area:	Lessons learned – Corporate knowledge building

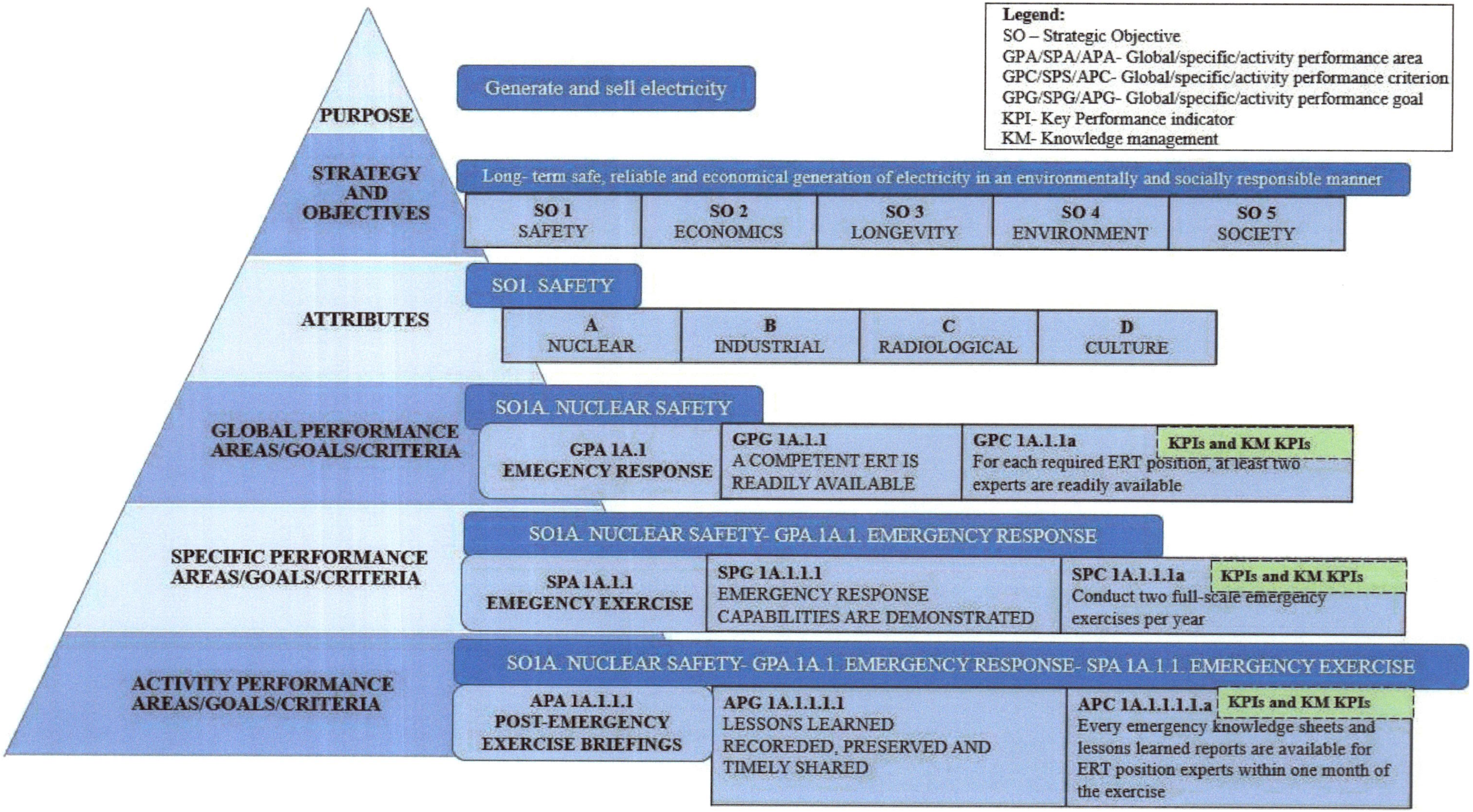

FIG. III-2. A specific example of layering performance criteria in determination of where KM KPIs and KM-related KPIs are used.

ANNEX IV.

KEY CONSIDERATIONS IN DEVELOPING PERFORMANCE INDICATORS

Typically, multiple sets of Key Performance Indicators (KPIs) support the measurement of KM and KM programme performance against a wide range of performance criteria that are applicable to KM and KM programmes. This may result in a large number of KPIs some of which are equal, the same or similar.

In order to find the best representative and optimized number of indicators, there are key considerations in the establishment of KM KPIs as to their characteristics and use that need to be taken into account during the KPI selection process.

Before going into details on the KPI selection and establishment methods and tools, it is important to reemphasize the few points for the measurement process and efforts for which the KPIs will be used:

— The purpose of measuring KM and KM programme effectiveness is to understand and identify actions to manage and improve the programme, its implementation and practice;
— Measuring KM and KM programme effectiveness is a part of the learning and informing culture towards continuous improvement and a sincere organizational commitment to safe and efficient business operations.

IV-1. FOCUS ON KNOWLEDGE MANAGEMENT PERFORMANCE

The KM performance measurement framework assesses the performance of KM activities or the overall KM programme and its application throughout the organization. As such, the KM KPIs need to solely focus on the KM and KM-related performance and the selection of the KPIs needs to consider that the indicators have the following characteristics:

— A clear link to the organizational strategy and objectives and cover the main areas of benefit addressed by the KM business case;
— Cover a range of benefits at different levels derived from the KM initiative itself (e.g. as a project) through to the organizational benefits to be realized from the initiative;
— Consider the need of the stakeholders (e.g. other business units/customers/the regulator/suppliers etc.) and their involvement in the KM initiative.

It needs to be kept in mind that KM KPIs directly measure the performance of a KM programme and the associated process and activities, as well as monitoring the KM and KM programme related performance of other programmes, processes and activities.

IV-2. APPLICABILITY AND ADAPTABILITY

There is a direct correlation between KPIs and the adequate representation of the performance area to be measured by the KPIs. As described in Sections 3 and 4, a KPI system is built on the purpose and performance areas (both global and KM) which are subject to revision in accordance with the applicable performance criteria as the business priorities evolve, morph, or are redefined based on the lifecycle needs of the business. Therefore, the main considerations include:

— Applicability to the stage and phase of the organization and the corresponding need (or priorities) for monitoring and measuring knowledge and relevant challenges. KPIs within the different phases of KM programme implementation are appropriately set on systematic way so that they are associated with the organizational lifecycle;
— Addressing both long and short-term needs (some KM initiatives may be short lived and eventually closed down while others may exist permanently throughout the lifetime of the organization);
— A holistic approach to change management in the organization concerning decisions, personnel, programmes, processes, activities, and roles and responsibilities both in decision making and execution.

IV-3. ABILITY TO MEASURE

As there is a direct correlation between the target for successful performance (i.e., performance criterion), it is important that KPIs can quantitatively or qualitatively measure and meaningfully indicate, the metric level of performance against a measurable performance criterion.

The selection of KPIs inevitably depends on the availability and reliability of processable data. However, the KPI's ability to measure relies on the methods and tools available to accurately calculate and properly represent the measured performance to be directly compared to a performance criterion (or several criteria).

IV-4. INFORMATIVENESS

The development and use of appropriate PIs helps employees at all levels to understand the implementation status and progress of KM-related activities and KM programmes for the most effective management of critical knowledge. KPIs also provide avenues to set/revise performance goals and expectations as well as gaps in the overall performance and help to communicate them to plant/site personnel.

In terms of the awareness to declined performance and actions to be taken to correct declined performance, maximizing the informativeness of KPIs therefore is important in two folds:

— KPIs provide simple and direct information about the level of success (or failure) of a certain performance (in this case the KM and KM programme performance) and the benefits achieved or not achieved;
— By identifying areas of both strength and weakness, the KPI assessment provides direction for developing reliable plans and forecasts for achieving goals.

Additionally, KPIs need to be informative enough to ensure that decision makers and programme authorities become, and remain, aware of actual practices, and values in the field, including those of external stakeholders.

An important consideration to characterize and determine the level of detail provided by KPIs, i.e. a KPI's granularity, is to accurately assess the type and extent of information to be considered in the decision-making process and communication strategy.

IV-5. MANAGEABILITY

Smaller sets of PIs naturally provide less information about all aspects of performance to decision-makers and are harder to communicate. More indicators consequently provide more information and details but may make the decision-making process, as well as communication more challenging owing to a potential overflow and complexity of indications. Furthermore, time and labour intensity (e.g. administration, maintenance and execution, etc.) increases proportionally (and sometimes even exponentially) with an increasing number of KPIs.

In order to ensure the manageability, there needs to be a degree of compromise to establish a set of KPIs that is adequate and sufficient, yet compact and conclusive, for a given circumstance. Naturally, fewer KPIs with more precise and meaningful indication for the decision making is preferred. A hierarchical approach with the underlying logic for effective and efficient coverage can help to achieve adequacy and completeness of the KPI. For example, the well-known Pareto principle (i.e. 20% of elements give 80% of total output) would give a good orientation on the prioritization of the available PIs.

To define the manageability of KM KPIs it is necessary to consider that regardless of the number or type of KPIs, the effectiveness of the KM programme can be measured by using clearly specified and described KPIs that have:

- A definition of what it is measuring and the reason;
- Input data and its processing;
- A grading scale and associated criteria/thresholds/rules, including the monitoring periods;
- A weighting and the basis of the weighting (e.g. importance to strategic, tactical, operational performance);
- Analysis and assessment methods and tools;
- Reporting requirements (as to role and responsibilities, scope and content, communication type, media and periodicity, etc.);
- A management review response and feedback process;
- A process of reporting and resolving adverse results and management feedback on corrective/preventive actions.

IV-6. SIMPLICITY

Simplicity means that the KPIs are not demanding and complex to understand and easy to obtain to ensure engagement from all stakeholders. The meaning of the KPIs, their collection, and calculation procedures need to be established considering the easiness of obtaining the input data (the necessary data are available or capable of being generated) as well as their consolidating, analysing, assessing, and reporting.

If there is a possibility for reaching the same (or similar) measurements, KPIs which can be collected, processed, and stored automatically need to be preferred. The ideal result here is to have a so-called ‘dashboard’, which automatically tracks and visualizes the levels and dynamics of the KPIs, as well as its non-complex reporting to the decision makers through algorithms.

IV-7. ACTIONABILITY

The key consideration for establishing KPIs is their contribution to previous decisions made and actions taken on the current and future performance status and hence, the KPIs are established with a primary focus on decision-making on any following actions. If a KPI cannot be viewed as a facilitator for KM-related decision-making it should not be selected.

As stated by the IAEA's International Nuclear Safety Advisory Group (INSAG):

> "Numerical measures must always be subject to careful interpretation and be used as part of an overall judgement about performance. They need to not be regarded as an end in themselves"[IV-1].

The KPIs are to be selected and tailored primarily towards identification of the underlying causes and precursors of any defect and deficiency (as well as the strength) in knowledge management, the KM programme and associated processes and procedures. This identification leads to the determination and implementation of corrective (or, in cases of strengths, cultivating) actions. The assessment of the KPIs can also be linked to the inefficiency of KM tools and technology, as well as issues related to interface and interaction communications of critical knowledge.

Accordingly, strategic, tactical, operational, and local actions can be determined based on their indication. Therefore, the consideration to be given to 'what will be done with the results of the indication' and 'what resulting actions are intended' requires answers to these questions:

— Will the target audience/user have control over the indicated status or condition?
— Will the KPI influence an action on the desired outcome?

IV-8. RELIABILITY AND REPRODUCIBILITY

The readability and reproducibility of the KPI is important to:

— Minimize or eliminate the susceptibility to manipulation;
— Minimize or eliminate the ambiguity;
— So, they can be validated;
— So that the accuracy of the data at each level can be subjected to quality control and verification.

IV-9. MAIN SELECTION CRITERIA

When selecting a KPI, it is important to consider the type of indication it will address. Additionally, utilizing the SMART approach to formulate the KPI can help ensure its success. Finally, it is crucial to evaluate how the KPIs will connect to existing KM and KM programmes within the organization to ensure seamless integration and maximum impact.

IV-9.1. Type of indication

Effective organizations develop and use metrics/indicators to monitor, measure, and improve KM and KM programme performance both reactively and proactively, by observing, collecting, analysing and assessing KPIs. The development of metrics for activities and processes could be undertaken in a manner similar to that described in INSAG-13 [IV-1]. For example, a set of indicators for KM and KM programmes may reactively (sometimes referred to as 'lagging') serve to measure the effectiveness of:

— The recent and current performance of KM-related tasks and KM programme implementation;
— The current awareness of past KM-related performance;
— The effectiveness of determination and implementation of preventive and corrective actions that have been taken to repair or improve the management of critical knowledge;

- — The attitudes and behaviour of staff, managers and authorities toward critical knowledge, KM and KM programme.

It is also important to establish and use forward looking (sometimes referred to as 'proactive' or 'leading') KPIs to inform on the awareness, identification and recognition of potential efforts to improve KM programmes, the associated processes and KM practices to conduct future activities. While such proactive measurements provide opportunities to anticipate, predict and pay attention, for example, to developing or accumulating issues, they indicate early signs of declining critical knowledge or KM performance, in order to take proactive measure to correct or change the path and trend that was leading towards inevitable failures.

The KPI can be either a quantitative or a quality measure. Developing and assessing qualitative KPIs, especially those focused on the future, can be challenging. For example, measurements of general or task specific personnel behaviour and attitudes in KM are typically qualitative in nature. Quantitative KPIs are easier to establish and provide a more practical input to the monitoring and assessment of the performance of activities with the observance of KM practices and initiatives.

Quantitative KPIs can be selected and established to measure the health of KM programmes and associated processes and procedures, by monitoring, for example the:

- — Number or consequences of errors in utilization or communication of critical knowledge and knowledge-based errors in the execution of an activity which may provide indications of:
  - o Lack of knowledge in the form of competencies and skills;
  - o Insufficient or deficient knowledge reflected in information management (such as analysis, work instructions, procedures);
  - o Insufficient or inadequate training to equip the staff and management with critical knowledge;
  - o Low value/importance of KM to staff and management;
- — Repeated incidents and deficiencies attributed to the lack of knowledge — when underlying errors and deficiencies could be quantified — which typically provide a measure/indication of:
  - o Programmatic failures or defects in the KM programme and corrective action programme (CAP) as to the determination and implementation of adequate, timely, or effective corrective actions to remediate the lack of knowledge after the earlier KM-based incidents/deficiencies;
  - o Organizational problems, such as not being a learning organization, particularly in such cases where the errors are the same as the ones that occurred in the past (internally or externally) due to lack of similar knowledge or KM activity;
  - o Issues with the correctness and adequacy of previous assessments or with analysis and communication of internal and external experience and lessons learned as knowledge acquisition or modernization (which may also point to issues, for example, with Operating Experience (OPEX) programmes and processes).

IV-9.2. Formulation Using the SMART Approach

Every KPI needs to be formulated in a way to fit all five characteristics of the classical SMART approach commonly described in the literature [IV-2], for example, as proposed:

— Specific: The KPI covers a specific performance area (or several specific areas) and compares it to specific performance criteria (i.e., there is a direct relationship between the indicator and area performance observed);
— Measurable: The KPI has a metric that is based on collectible data or can be qualitatively scaled or analysed;
— Achievable: The KPI is linked to a realistic outcome (e.g., to action(s) or decision(s) that can be taken);
— Relevant: The KPI is relevant (or related) to a KM performance area and criterion that is aligned with the strategic business objectives and KM strategy (i.e., it is capable of indicating the performance level of the observed area);
— Time-bound: The KPI covers and represents an explicitly defined time period.

IV-9.3. Connection to the Existing Processes

Knowledge Management is a holistic, crosscutting and overarching process that predominantly pulls on and aligns existing company tools and processes and, consequently, KM and KM programmes are (or need to be) fully integrated into Integrated Management System (IMS) programmes and processes that already exist. Therefore, at the starting point of the KM KPI selection process, a particular selection criterion is to be used, as many existing KPIs applicable to the performance measurement of KM and the effectiveness of the KM programme. Reviewing all KPIs that have been previously used in the organizations, programmes, processes, and procedures (e.g. there may already be existing KM-related KPIs for the other programmes and processes within the plan/project organization) helps to avoid duplication.

The existing KPIs (either for the other programmes and processes within the organization or already established KPIs for KM or KM programmes by peers in the industry) could also be adopted/derived and used to measure KM and KM programme performance, if the existing KPIs or a link/derivative of those are adequate and sufficient to measure against the applicable performance criteria. For example, there might be an existing KPI used by the plant/project organization for measuring the effectiveness of maintenance programme performance, say: 'actual versus expected maintenance duration', which can be derived into a KM/KM programme KPI as 'outage extension due to delays caused by unavailability of maintenance experts.

Some organizations may need to measure other aspects and create a new KPI indicator that is more suitable and applicable to their specific conditions, strategies and areas of interest and attention to help with continual improvement. Introduction of new PIs is justified only if there is an obvious gap in the coverage of key KM and KM programme areas that are uniquely important for an organization.

IV-10. SETTING VALUES AND METHODS

Key Performance Indicators need to include a predetermined set of targets, a means of measuring current activity, a means of comparing current activity with each target and an associated value. KM KPIs need to be categorized based on contribution, use point system, and provide a feedback loop for correction.

IV-10.1. Determination of Metrics and Tools for Monitoring and Assessment

The effective measurement of the implementation and improvement of KM processes and practices requires the monitoring and assessment of existing performance to guide the organization to ensure the successful implementation of KM and KM programmes in future phases. To do this effectively, a KPI system needs to include some important elements before evaluation may be fully established. These elements are usually a predetermined set of targets, a means of measuring current activity, a means of comparing current activity with each target and a means of correcting deviations from the targets.

Once the set of KPIs are selected, there needs to be an associated value that will be compared to the associated performance criteria with determined metrics, assessment criteria, rules and methods for each performance indicator. These KPI values can be in different forms, rules and methods.

IV-10.2. Examples of Key Performance Indicator Values

A confirmative condition that measures the performance by occurrence or existence of a specific condition described in the performance criteria, such as the measurement of performance by a checklist that will compare the indicated condition with a criterion and confirming success by a 'yes/no' to the satisfactory performance criterion for example, the KM performance that mandates:

- — A KM policy exists;
- — Full-scale emergency drills take place once a month.

A numerical value that will be deterministically or statistically calculated (over a range or discrete) and then compared to the performance criterion, which is a numerical value, for example:

- — Full-scale drills take place once a month with new team members shadowing experienced members.

Numerical value could be confirmed by keeping track of the participation of new team members to the drills, such as: Percent of drills in which new team member for Position X participated shadowing the expert team member.

A judgment value that will indicate characteristics of a decision or opinion, such as:

- — Awareness of KM and KM programme by employees;
- — View of managers on the application of KM and KM programmes.

Judgment value could be based on the collection of a numerical data (e.g. employee survey, or management observation cards), however interpretation and the process success is judged based on the performance criterion which is an expectation.

Typically, the majority of the KM KPIs are expected to be compared, presented and assessed numerically which necessitates setting a clear structure of data and data source, description of how to create and process the data and methods to calculate the indicator, including regression and mathematical rules, criteria/thresholds and methods of evaluation and interpretation of assessment results.

In setting the metrics and methods, it is important to note that:

- — The indicators and methods are not susceptible to manipulation;

— The metrics are meaningful as to the performance criterion comparison;
— The calculation and comparison of indicators with the performance criterion serve as part of an overall judgement about performance and are not the end product but rather the starter of the process with an outcome of improvement and follow up actions.

Moreover, numerical measures (e.g. indicators and metrics) for measuring KM programme effectiveness are set:

— To understand and to interpret the status and trends of corporate critical knowledge and its management and to act upon, when necessary, for correction and improvement;
— Are not for rewarding or punishing individuals or organizations;
— Are not for 'bean-counting' events/errors/failures, unless their occurrences and trends are being evaluated and decisions were made to take corrective or preventive action as a result of performance monitoring.

The KPI value setting process may also include a set of possible actions to be taken at all levels based on the outcome of the assessment of each indicator, including the trigger/threshold value for actions and action levels.

IV-10.3. Determination of Weighting and Scaling

One of the aspects of KPIs is that their significance needs to be understood throughout the organization, particularly by those who are establishing and managing the KPIs. The first step is to assign a significance value to serve the overall and specific performance requirements and expectations in the order of (from the highest significance to lowest):

— Purpose of the business;
— Strategic objectives;
— KM programme goals;
— Divisional, departmental, sectional performance expectations;
— Personal performance expectations.

In many organizations, some KPIs prevail over others (for example, safety over economics, financial over non-financial, operational over strategic, individual over team, etc.). Of course, the composition of indicators depends on the characteristics of the organization, its strategy, and goals, but in any case, it is important to remember and try to maintain a difference in significance between the different types of indicators, particularly for the impact and influence of decisions that are made as a result of the KPIs.

For example, KM KPIs may categorize 'nuclear safety events and close calls resulting from critical knowledge related reasons' based on the severity or consequence of the impact on strategic business objectives and global performance area/criteria. A point system is then used in conjunction with the level of significance to be presented to management on a periodic basis, as well as a rolling average, which is used as an indicator for measuring the overall number and significance of KM related events to take timely action.

IV-10.4. Reverification

The last step in the establishment of KM KPIs needs to also provide a feedback loop for the step-by-step process discussed in Sections 3 and 4, matching the final KPIs to business strategy and objectives, as well as the corporate values and expectations for knowledge, to check the process for correction or fine tuning.

An integrated set of KPIs that are linked to strategic objectives can provide a meaningful and comprehensive evaluation of the effectiveness of the KM programme for the organization's overall performance. It needs to be noted that:

- An organization can use multiple PIs to measure the effectiveness of its KM towards achieving a strategic objective, or conversely, multiple strategic objectives can be linked to a particular KM PI (Fig. III-2 in the Annex III is an example of linking a particular KM PI to multiple strategic objectives) and vice versa;
- An organization can use same, similar or different specific KPIs linked strategic tactical or operational objectives and performance expectations, while KPIs are reviewed and acted upon at the authorization, coordination or execution levels, as discussed in the introduction to Section 4.

Figure IV-1 illustrates this reverification by another example of aligning overall strategic business objectives of a specific nuclear facility at a certain phase of its lifecycle (in this example, an NPP at operation phase) with KM and KM programme goals and associated indicators.

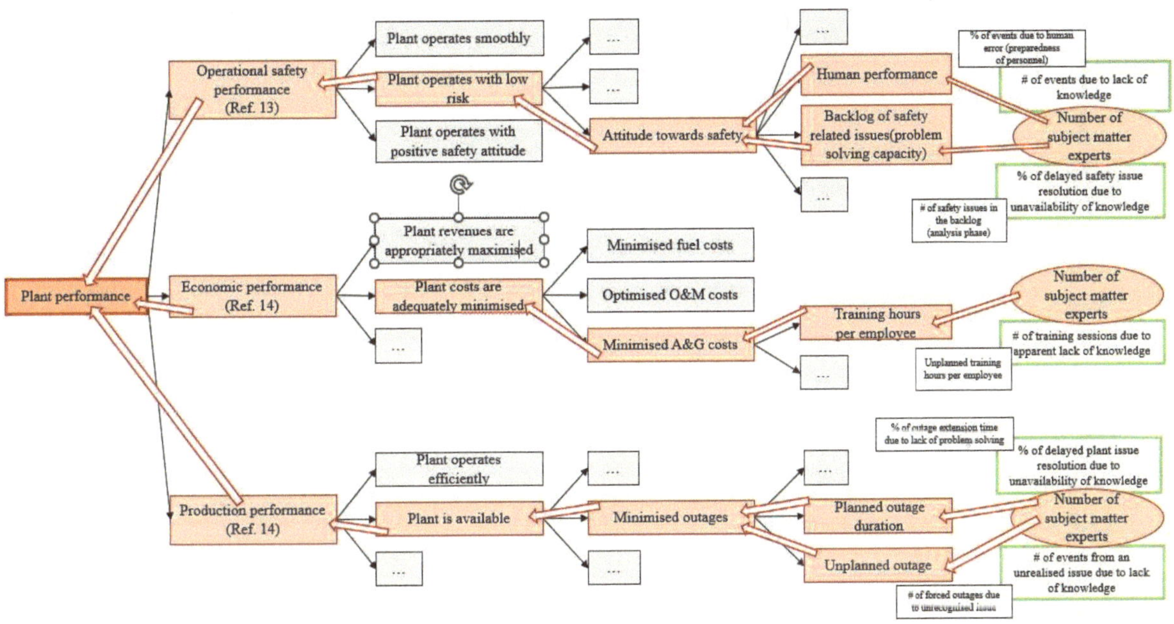

FIG. IV-1. An example of the alignment between the overall plant performance and strategic objectives and KM programme goals linked with KM performance standards and associated indicators (think arrows) and vice versa (thick arrows). O&M — operation and maintenance, A&G — administrative and general.

REFERENCES

[IV-1] INTERNATIONAL NUCLEAR SAFETY ADVISORY GROUP, Management of Operational Safety in Nuclear Power Plants, INSAG-13, IAEA, Vienna (1999).

[IV-2] DORAN, G. T., "There's a S.M.A.R.T. way to write management's goals and objectives". Management Review. 70 (11): 35–36 (1981).

ANNEX V.

MEASURING PERFORMANCE OF KNOWLEDGE MANAGEMENT ACTIVITIES IN DIFFERENT PHASES OF ORGANIZATION LIFECYCLE

During the implementation and plant realization phases, it is important to track KM performance and the key indicators to ensure the successful execution of the KM programme. In the design phase, specific KM performance and key indicators need to be established to guide decision-making and monitor progress. Similarly, during the construction and commissioning phases, unique KM performance and key indicators need to be identified to optimize KM programme outcomes. Effective KM during the operation phase can significantly impact the success of the decommissioning phase. This includes maintaining accurate records and documentation, as well as ensuring proper training and knowledge transfer to those involved in the decommissioning process. Ultimately, a comprehensive KM strategy throughout the entire lifecycle of a plant can led to improved efficiency, reduced costs, and increased safety.

V-1. PRE-PROJECT PHASE

It primarily includes the processes and activities that enable a knowledgeable decision to be made on whether to launch a nuclear programme for power generation.

Typically, in a newcomer country, the KM aspects during the pre-project phase have two parts:

- KM during the current performance of tasks will primarily focus on the knowledge lifecycle elements of identification and acquisition and will mainly be concerned about the people/human and organization/process components of knowledge management. For a newcomer country building their first nuclear power plant, it is very unlikely that a KM programme exists; however, it would be a good practice to actively conceptualize and apply knowledge management, invest in education programmes, develop national networks, preserve and transfer the knowledge being identified as useful for the current and next phases;
- As a specific performance area of the pre-project phase, human resource development assessments and evaluation "describe the knowledge, skills and attitudes of the multiple disciplines required for a nuclear power programme and a strategy for obtaining and maintaining the needed personnel" [V-1]. This is the foundation of the KM strategy for a potential nuclear project and eventual plant and the starting point of a KM development roadmap and initialization of a KM programme for introduction later in the project development phase.

In both parts, the performance areas and criteria of this KM process first need to be linked to the current and future strategies and strategic objectives of the organization, both for the overall programme/project and for the knowledge.

The performance of the pre-project phase activities requires technical, commercial, financial, and legal expertise with a certain level of knowledge in a given specific performance area, for example, knowledge on international law and regulations on nuclear safety, security, and safeguards for the activities in the specific performance areas of legislative, regulatory, and safeguards frameworks.

During this phase it is important to identify the full range of disciplines, as well as the need for specialized recruiting, education, and training that will be needed for the nuclear energy programme.

The level of knowledge to perform pre-project activities depends on the time needed to complete a comprehensive study and to establish recommendations for deciding on whether or not to launch a programme. This time frame depends on the level of expertise, experience, and knowledge of the decision makers. Therefore, the decision makers benefit from the engagement of international support (such as, the IAEA, foreign consultants and subject matter experts), as well as domestic experts (whose non-nuclear knowledge would be applicable to produce detailed evaluations and assessments) to produce the comprehensive study within an expected time frame. For example, a performance criterion may mandate to:

> Within one year, provide the electrical grid report, prepared by highly knowledgeable and experienced grid planning experts, comprehensibly describing the electrical grid size, configuration and reliability necessary to accommodate the addition of an appropriately sized nuclear power plant and the likely extent and cost of grid enhancements that will be needed. Ensure the validity or make revisions six months before the decision on launching a nuclear programme.

Even if the country initially relies extensively on knowledge and skills from other countries, it needs to consider how to develop its own long-term knowledge and skills.

For example, a KM performance criterion (presuming a KM policy to acquire and maintain critical knowledge exists within the entity) could be:

> Ensure and maintain availability of at least two highly knowledgeable (i.e., who have prepared at least two electrical grid reports for similar projects accommodating nuclear or other types of large-size electricity generating units) local grid planning experts, one for preparation and one for independent review, until the decision is made on launching a nuclear programme.
>
> In case no qualified local experts are available, hire and maintain two qualified international experts during the pre-project phase and ensure acquisition of foreign expert's explicit knowledge by local experts working together with them through quarterly report progress review and knowledge exchange meetings.

For example, the purpose of the entity at the pre-project phase could be to:

> Provide a case and recommendation for a national decision to undertake (or not undertake) a nuclear power programme.

This purpose may have the objective to:

> Have a clear understanding of conditions and options for all infrastructure issues, as well as the needs, recommendations for the strategy and policy directions including plans and budgets for development and implementation, in order to make a knowledgeable decision for launching a programme and for developing a project for a viable, safe and economical nuclear electricity generation in the future national energy mix with a tolerable risk.

These objectives would then necessitate:

— The identification and assessment of the needs, benefits, gaps, risks, and consequences of launching a programme;

— The identification and description of strategies to develop the programme including those to fill all identified gaps in programme development;
— The identification of strategies to eliminate or minimize all identified actual and potential adverse conditions and risks as reasonably practicable.

Among others, these objectives are selected to illustrate the activities that are directly knowledge-based. These global performance areas, based on the strategic objectives, can be decomposed into specific performance areas.

Accordingly, the performance of associated pre-project phase activities requires a combination of technical, commercial, financial, legal, and educational expertise with knowledge for specific performance areas. It also involves integrated performance by a large number of organizations (private and public government controlled, national, foreign or non-governmental international). For example, in the legislative and regulatory frameworks and safeguards, experts with knowledge of international laws and regulations of nuclear safety, security, and safeguards are needed. In the electrical grid performance technical experts, such as electrical engineers with knowledge on the existing electrical grid system and on short- and long-term grid planning are needed.

The overall manpower need for that stage is relatively modest, mostly oriented at directing, coordinating, and registering data.

A nominal time frame of about 1–3 years may be planned for completing Phase 1. This time frame could be shorter or longer depending on the resources provided, including the expertise and competency available to initiate and perform activities in global and specific performance areas.

One specific performance area in the pre-project phase, namely human resource development, has a high importance for knowledge management. Describing the knowledge, skills, and attitudes of the multiple disciplines required for a nuclear power programme and a strategy for obtaining and maintaining the needed personnel, is the foundation of systematic KM and initiation of a KM programme. In this area, which knowledge is essential and why as well as when and where it needs to be available (e.g., its acquisition, retention, etc.) for achieving specific programme/project strategic objectives and required/expected programme/project implementation performance, is defined. This forms the knowledge value strategy and vision of the organization, based on consideration of:

— The significance, urgency, or uniqueness of the programme and project decision and of the knowledge needed for that decision;
— Having and maintaining adequate and appropriate organizational capability and competency;
— Balancing the projected and planned financial costs associated with obtaining (purchasing) knowledge from the external sources and maintaining the knowledge in-house, including the value/impact of owning the knowledge.

V-1.1. Project Development Phase

After a decision is made to launch a nuclear power generation programme, the next phase to develop the project for new (or additional) nuclear power generation consists of activities associated with the stages to invite and evaluate bids to decide on a contract to design and build the nuclear plant. Activities in this phase typically include the collection and determination of the project input, such as legislative, legal, and regulatory requirements, codes, standards and specifications, etc., all of which require a degree of critical knowledge to perform. The project preparation activities require a combination of technical, commercial, financial, and legal expertise with particular knowledge, for example, of electrical and nuclear technologies, safety and quality assurance, techno-, power- and environmental-economics and laws, as well as major project financing.

At this stage, knowledge is needed to serve the objectives of:

— Building up the project requirements and expectations, including a knowledgeable workforce;
— Making a selection for a contract to design and build the nuclear plant.

Furthermore, expert knowledge is needed during by the project owner entity at the later stages to understand and be aware of design, construction, and commissioning activities and associated issues to be able to make informed decisions on its acceptance and approval.

Some of the needed knowledge (including the non-nuclear knowledge that is applicable to a nuclear project) exist as a permeated resource within several organizations working together towards the project development. It is critical to have competent staff being aware of and/or trained in available state of the art models, methods, and techniques, as well as being informed about experience and knowledge of others in conducing of analyses, evaluations, and assessments. At this stage, the project owner may benefit from the engaging international consultants and subject matter experts to coordinate the project specifications.

Therefore, in a newcomer country the management of knowledge mainly deals with all three components of KM, namely the people/humans, organization/process and technology/tools. For example, the purpose and strategic objectives of the entity at this stage could be to:

> Develop and plan the project for a safe and constructable NPP, with appropriate technology, location, and reasonable financial and scheduling risks, fully complying with all applicable laws and regulations, to be maintained and operated by domestic resources.

It need sto be noted that, in the case of an existing and experienced owner/operating organization planning a new or additional power plant project, existing knowledge, and experience includes not only the lessons learnt, but also the specifications and performance of existing technologies and products, as well as some of the vendors' management systems, behaviours, and values.

Once the specific performance area and criteria are defined and agreed by all stakeholders participating and interfacing in project development, common or specific KPIs can be developed in a holistic and integrated manner. Performance area of HRD planning will be applicable to different stakeholders (such as the academia, industry, government, utility, etc.).

In this phase the knowledge and skills needed in next phases as well as staff requirements, recruitment, and training plans based on capacity gap analyses of the involved organizations need to be identified. The plans usually cover competency, education/training, knowledge, and experience expectations and requirements based on the KM strategy of the decision maker and need to consider bilateral and international competence building activities for information and knowledge acquisition.

V-1.2. Project Implementation and Plant Realization

Early in this phase, the plant owner/operator assumes a key (and gradually expanding) role in the development and implementation of the NPP project. The owner/operating organization's need for knowledge in order to make project-related decisions gradually increases as the design is developed (design stage), implemented (construction stage), and verified (commissioning stage).

These internal knowledgeable experts inform/advise the decision makers in the owner/operating organization to enable the project to progress with safety, technical adequacy, schedule, and budget. The owner/operating organization also solicits and acquires knowledge from independent external experts in both highly specialized or trivial issues.

This is the time to record information and explicit knowledge, for example design basis, approaches and methods, to be available for later needs and activities during this and subsequent phases of the NPP's lifetime. This is a critical KM aspect, as the acquired information, knowledge, and records of those, in accordance with the strategic business objectives, and established strategy for institutional knowledge in operation and decommissioning phases, form the pillars for:

- Providing, maintaining, and improving institutional knowledge through competent internal expertise;
- Exchanging accurate and pertinent information and knowledge with external organizations;
- Establishing, within KM programme, key knowledge areas, such as:
- Technical and scientific knowledge on design, construction, and commissioning from in-house and external technical stakeholders (e.g., responsible designers, vendors, contractors, external consultants, and experts);
- Application of programmes, processes and procedures, and associated administrative controls (e.g., industrial and nuclear safety, safety, project management, component manufacturing, equipment supply/procurement, quality assurance, training, document control, problem identification and resolution, performance improvement, etc.).

The strategic goals of the project owner/operating organization to provide its staff, particularly those who will be part of the plant personnel later during the operation phase, are to:

- Closely follow the progress of design and construction activities such that the technical staff is aware of and attentive to design and configurational objectives and fundamentals of the plant, for example design philosophy, basis, approach and methods, structures, and systems;
- Closely shadow and observe the activities performed by the external organizations, who hold and apply their knowledge during construction and commissioning, in order gain information and knowledge on facility configuration and programmes, processes, and procedures;
- Stay up to date with the developments in the nuclear industry and use operating experience to expand institutional knowledge by being aware of best practices and/or pitfalls, as well as the advancement of state-of-the-art knowledge.

It needs to be emphasized that knowledge transfer during project implementation is one of the most important activities making KM and KM programme the highest priorities of specific performances. Through knowledge and information transfer and recording, the owner/operating organization can take over the ownership, responsibility, and accountability of the plant and establish a meticulous KM strategy for its strategic goals, requirements, and expectations. As such, this is a critical phase of KM since it forms the future NPP knowledge base for its safe, reliable, and economical operation, maintenance, and modification. Therefore, monitoring needs to ensure that knowledge transfer (i.e. the acquisition and maintenance of knowledge about the initial design and configuration of the plant) is highly effective and its success measured by KPIs.

KM specific performance areas/criteria and associated KPIs in this phase of organization lifecycle need to be related to:

- Acquiring and recording all necessary design and construction related critical information (e.g., design, its basis and changes to it) during design realization (i.e., construction), licensing and validation (i.e. commissioning) that are obtained and recorded in programmes, processes and procedures;
- Setting up long-term plans for information and KM of design and its resources, including the determination of needs for, and establishment of, a sustained knowledgeable and well-trained staff (for example, building up groups of technical staff who can perpetually maintain the design and configuration integrity of the facility, as well as preserve and transfer the design information and knowledge to the next generation of staff for the lifetime of the facility);
- Building up adequate knowledge of in-house staff (individual and group of experts)

V-1.3. Common Knowledge Management Performance and Key Indicators During Project Implementation/Plant Realization

KM performance area that can be monitored and measured with commonly applicable KPIs at different stages of project implementation, i.e., design, construction, and commissioning, are provided below.

Recorded institutional knowledge: As large volume and essential information or knowledge is provided, observed, collected, and gained during the project implementation, they are recorded and maintained for critical knowledge needs that could occur later in the project/plant lifetime. The information and knowledge gained on the plant design, construction, and commissioning (e.g., design and its basis, equipment and system as-built conditions, test results and performance trends, critical operating and design margins, important technical and analytical observations, etc.) are recorded and maintained to the maximum extent possible. Similar KM KPIs monitor and measure the collection and retention of:

- Information beyond the regulatory requirements (e.g., those required record keeping under an integrated management system that prescribe record control for quality assurance purpose, etc.);
- Explicit and tacit blueprint and physical facility knowledge.

This is accomplished through a formal and systematic record keeping process for information and knowledge that is adequate, timely, as clear and comprehensive as possible and in accordance with the corporate strategy for current and future state of institutional knowledge.

It is also important to acknowledge that there are informal mechanisms for knowledge acquisition and transfer during design, construction, and commissioning, which may vary widely (such as prompt verbal feedback to line colleagues and leaders, discussions between internal and external points of contact, occasional discussions within groups and senior management, etc.). Such informal mechanisms need to be considered as KM tools and methods and the performance in their identification, recording and preservation as a part of institutional knowledge needs to be monitored and measured and hence, would require associated specific performance criteria and KPIs.

Training: Training enables individuals and organizations to understand and perform the required tasks with thorough knowledge of fundamentals. The training programmes for knowledge gain/transfer and utilization use a wide variety of standard training methods (e.g., formal professional education, specific classroom lectures, on the job training (OJT), on the job participation (OJP) such as expert shadowing or human resource sharing in a specific activity, etc.). Fortunately, during the project implementation stage, there are abundant opportunities to gain knowledge that include:

- Broad discipline education (duration ranging from several months to a couple of years with large class size) for gaining information and knowledge on overall and specific fundamentals of design and construction;
- Specific discipline classroom training (CRT) (limited class size and duration ranging from several days to weeks) for gaining knowledge on specific activities performed for design, construction, commissioning as to its purpose, scope, performance and associated fundamentals, applicable tools/methods, as well as associated programmes, processes, and procedures;
- OJT by individual work assignments under expert supervision in an actual activity in design, construction, commissioning area, discipline or organization to gain specific knowledge and experience, including human resource sharing/embedding arrangements;
- OJP by expert mentor-protégé performance of tasks in design, construction, commissioning area, discipline and organization activities, or shadowing and observing expert performance of tasks, for gaining in depth experience and knowledge.

This will be accomplished through formal and systematic training programmes and activities that are adequate, timely, as clear and comprehensive as possible, and aligned with the corporate strategy for current and future state of institutional knowledge. Accordingly, KM KPIs measure and monitor the performance of critical knowledge capture and its effective use consistent with the corporate strategy for current and future state of institutional knowledge.

Information Management: Documentation to be transferred consists of all technical information produced for the design, licensing, construction, start-up, and operation of the plant. It contains design documents and drawings, various background documents that form the basis for the design and project management documentation including programmes and procedures, such as quality assurance and design control. Through training on preparation, technical content, and the use of documents specifically applicable to the actual design, newcomer engineers become familiar with the design process and procedures necessary for control of engineering activity and interfaces.

If in this stage mature KM programme is already established and implemented, the KPIs to measure the effectiveness of establishment, communication, and application of short- and long-term KM strategy and policy would also need to be developed.

V-1.4. Design Phase Specific Knowledge Management Performance and Key Indicators

At the design stage of project implementation, the responsible designers (i.e., the technology owners) manage the design activities based on their own institutional knowledge, internal processes, requirements, and competencies. However, the project owner, bears the responsibility to:

- Approve the NPP design, as it develops including its features and necessary changes;
- Ensure that the design and physical facility meet or exceed the legal, environmental, safety, contractual, and other technical and regulatory requirements, as well as economic and efficient plant performance needs and expectations;
- Obtain regulatory approval for construction and operation (i.e., construction and operation licenses) with full presentation of the safety case to the regulatory authority with adequate design documentation.

Therefore, there is a need for critical design knowledge to gain adequate awareness and understanding of the design, licensing, and associated issues and their resolutions for taking decisions on the design acceptance and approval. There also needs to be sufficient knowledge available to identify and resolve any exceptions, deviations, and substitutions to project design input document and project schedule progress and milestones, as well as to manage the interface with the regulatory body on the design and the safety case involving critical licensing knowledge.

Depending on the KM strategy and activities established and applied during the project preparation phase, some of this needed critical knowledge could exist in the project owner's organization (either internal or external staff). Alternatively, the knowledge of external entities (international consultants and subject matter experts) may be used. In the first case, the already existing KM strategy and KM programme may dictate to establish, expand, and maintain the design and licensing knowledge of staff in design input, models, methods, and tools and to collect, record and preserve design information (e.g., design basis, analyses, evaluations, assessments, drawings). In the latter case, the KM strategy and KM programme would aim to transfer knowledge from the external entities supporting the project for reviewing and overseeing design and licensing work. This knowledge transfer goal includes the acquisition/transfer, collection and compilation of design information and knowledge from the responsible designer.

Accordingly, KPIs need to be established to measure and monitor these KM performance areas and criteria (as they relate to the business strategic objectives and global standards of performance/criteria).

For example, members of the owner's technical and scientific staff need to be closely involved in the design activities performed by the external entities who have the knowledge in order to gain in-house knowledge and capabilities, e.g. through job shadowing programmes **Error! Reference source not found.**. Also, senior experts of the owner/operating organization need to maintain close contact with the responsible designers and the vendors who are working in the construction and commissioning to ensure that the necessary knowledge is being transferred. Therefore, in accordance with this KM performance area, it is beneficial to establish KPIs to measure and monitor the performance of transfer of knowledge.

It needs to be noted that, as a part of information and knowledge acquisition and utilization, understanding and use of the computer codes and other tools used in the design is an important aspect. Typically, there are several hundreds of computer codes (some of which are proprietary) used in core design, system design, plant performance, and safety analyses, structural analyses component design, and manufacturing and special knowledge is needed in use and maintenance of such codes along with their documentation.

V-1.5. Construction Phase Specific Knowledge Management Performance and Key Indicators

In the construction phase, the extent of project owner's involvement increases and extends to the erection, installation, and assembly of the plant. Particularly:

— The project owner/operating organization oversees, reviews, and approves the work performed by the organizations that build or supervise the field implementation (including the design changes and physical modifications). It is therefore necessary that decisions are driven by knowledge and expertise, and the project owner/operating organization needs to be the knowledgeable customer gradually taking over the facility. The project owner/operating organization needs to have the knowledge for verification and assessment of progressing design and realization of the plant, as well as for managing the authorization of operating organization.

— The plant is gradually assembled and there is open and progressive access to all parts of the facility. This provides opportunities to see, record (explicitly and tacitly), and understand the physical configuration of the plant. This builds up the physical *plant* knowledge which will become very important during the later phases of plant lifetime, operation, and decommissioning, in addition to the design knowledge and licensing knowledge.

— The knowledge gained in construction, installation and assembly techniques, methods, and tools, including those in work management, scheduling, project management are very beneficial during the operation phase. This knowledge will be important for physical facility changes during operation and maintenance of the plant, particularly for very extensive and complex changes to the operating facility, such as refurbishment, major equipment replacement, major structural changes/additions, which in some cases could be viewed as 'reconstruction' of the facility.

Depending on the KM strategy and activities established and applied in the project preparation phase, some of needed knowledge could already exist in the project owner's organization or within external organizations. Already existing KM strategy and KM programme may dictate to establish, expand, and maintain the plant knowledge about the physical setup and composition, the KM strategy and KM programme would also aim to transfer knowledge from the external organizations supporting the project for reviewing and overseeing design and licensing work. This knowledge transfer goal needs to include the acquisition/transfer, collection, and compilation of knowledge from the responsible designer, architect/engineer (A/E), and other parties that participate in the design and construction. KPIs are established to measure and monitor KM performance areas and criteria related to the business strategic objectives and global standards of performance/criteria in the performance areas, such as:

— Safety (particularly industrial safety during construction);
— Quality (control and assurance for design and construction);
— Schedule;
— Costs;
— Human resource development.

Another performance standard could also specify that the owner's staff need to be closely involved in the design and construction activities to gain in-house knowledge and capabilities, such as, shadowing/embedding the internal in-house staff in the activity and task performance.

V-1.6. Commissioning phase Specific Knowledge Management Performance and Key Indicators

The commissioning phase is the last stage before the operation (noting that construction and commissioning phases, in most projects will overlap). During commissioning, the functionality and operability of Systems, Structures, Components (SSCs) are verified by the tests performed on SSCs as a part of design validation. Consequently, the plant SSCs will be handed over for operation and the responsibilities for, and authority over to the operating organization. Plant programmes, processes, and procedures are transferred from the design and construction organizations to the commissioning organization and to the plant owner/operating organization.

In the commissioning stage is possible to observe and understand the SSCs status, operation, and performance under operating conditions and the last possibility to get close to most of the plant SSCs without any radiation exposure for personnel before the plant enters the operating stage. As the activities also depict human machine interface, they help understand the ways the facility layout ensures ease of accessibility to the SSCs for their inspection, surveillance, and maintenance.

The following aspects related to KM need to be taken into account at this stage:

- When the commissioning activities are conducted, a level of knowledge is needed by the project/plant owner/operating organization to a review of and approve the test results to ensure that the design and physical facility meet the requirements and expectations of the strategic business objectives, e.g., safe, economic, and efficient plant performance;
- Commissioning activities provide a good opportunity to gain baseline operation knowledge and plant knowledge, particularly for systems and components and operation knowledge, as well as for the programmatic knowledge on plant programmes, processes, and procedures, particularly those for operation, maintenance, testing and surveillance, problem identification and resolution.

Furthermore, the OPEX shows that a formal, structured, and effective facilitation is essential not only for the transfer of responsibilities and information from the construction teams to the commissioning teams and, then on, to owner/operating organization's operation staff, but also for the knowledge gained by experience, observations and lessons learned from the tests. This knowledge transfer is accomplished by a mature and effective KM programme as the owner/operating organization's capabilities and competencies are nearly in its complete role for the plant operation and maintenance.

Accordingly, the effective performance of KM and KM programme during the commissioning phase will involve the knowledge lifecycle elements:

- Identification;
- Collection;
- Acquisition;
- Preservation;
- Utilization.

And will deal with all three components of KM, with the people/human, organization/process, and technology/tools. Performance of these KM activities for the management of knowledge will accordingly be linked to the overall plant operation and maintenance strategic objectives of the decision-making party (for/of the plant owner/operating organization) and the specific KM strategy and objectives.

The strategic business objective is to complete commissioning tests safely, correctly, properly, and timely. The responsibility for the work is generally assigned to the teams of designers, suppliers, constructors. However, in one moment responsibility for plant operation will be turnover to plant own staff. One of the business strategic objectives and associated KM strategy is to establish and maintain in-house expertise. For that purpose, plant own staff needs to be a part of the commissioning test team. This combination ensures that explicit and tacit knowledge is gained/transferred. At the same time, it also ensures that the right expertise is made available in a timely way and the defects and learnings concerning the future operation of systems or component are identified, resolved, and recorded. Furthermore, on that way, the explicit and tacit knowledge, on system's operation is properly incorporated into plant documents, such as the operation instructions and procedures including access for manual operations.

V-1.7. Plant Utilization (Operation) Knowledge Management Performance Key Indicators

Once authorized to operate the NPP (i.e., operating license) the owner/operating organization is fully responsible for its safety [V-2], [V-3] and for taking safe, reliable, and sound operational decisions. Together with this responsibility, the owner/operating organization also upholds efficient and effective utilization and performance of their NPP. To accomplish this, the NPP owner/operating organization needs to consider, evaluate, initiate, perform, implement, or manage plant activities and assets. These actions require making numerous decisions on plant activities and assets to achieve the purpose and objectives of business, i.e., safe, regular, and efficient production of electricity and energy, particularly on maintaining/increasing:

— Safety and operational margins;
— Plant performance;
— Return-on-investment.

These decisions need to be made in an informed manner with consideration of all relevant information provided to the decision makers based on the knowledge, proficiency, and competency by the supporting individuals and organizations, within and outside the owner/operating organization. This sound and timely information and knowledge that is necessary to support right decisions need to be available and adequate, which requires an effective and efficient KM strategy managed by an established, mature, and structured KM programme that is successfully supporting the decision-making process. As discussed earlier, a key process to ensure successful and continuously improving performance of KM and KM programme is to measure and monitor the performance of KM, KM programme and all associated processes and procedures by established indicators that demonstrate the strengths and weaknesses. There may be variations in the KM strategy and KM programme focus — and hence, KM performance areas, standards of performance, performance criteria, and associated PIs — during plant utilization, depending on the plant activities, as well as the corporate strategy, style, and tradition of owner/operating organization.

However, there is fundamental knowledge that is important for decision making on plant utilization and that needs to me managed, including:

— Operation knowledge;
— Plant design knowledge;
— Maintenance knowledge;
— Technical and scientific (engineering) knowledge;
— Licensing knowledge.

Additionally, interfaces and interactions between organizations and between individuals occur at all times of plant operations so that departments and people need to be collectively and continuously aware and vigilant of possible acute, latent, and/or cross effects of issues related to institutional knowledge within overall owner/operating organization. Therefore, an effective KM and KM programme during operations requires application of the systematic management process to all activities at the plant/site, e.g., operation, engineering, work planning, oversight, surveillance, testing, chemistry control, radiological control. This also points to another important knowledge area namely administration and coordination knowledge of plant programmes and processes, which is essential to control and manage overall plant performance.

Noting that knowledge is not only utilized but also gained, shared, and updated continuously during the operation phase and it is, among others, preserved and transferred, the KM and KM programme performance includes all knowledge lifecycle elements and concerns all three components of KM.

Although the common KM and KM programme performance and KPIs apply throughout the plant operation phase (including what is presented in Section V-1.3), there are stages during the operation phase where specific KM elements and components and associated KPIs are applicable (or common KPIs are weighed differently). Particularly, as a NPP gets closer to the end of its operating license term, the owner/operating organization decides on ceasing operation or extending the license period. Consequently, extended operation stage and/or operation stage in transition to decommissioning are distinctive within the plant operation phase. Therefore, it is prudent to discuss them separately, for the purpose of this publication, as:

— Operation (i.e. plant operation as initially purposed and established within originally licensed period;
— Long-term operation (i.e., plant operation beyond the original licensed period with extension of license);
— Operation in transition to decommissioning (i.e. plant operation while preparing for plant disposal).

The following sections discusses specific KM and KM programme performance areas, standards of performance/criteria during these stages of plant utilization phase.

V-1.8. Operation

Generally, the primary purpose of NPP utilization is to generate electricity safely, reliably, and economically. It defines the NPP owner/operating organization's strategic business and global performance categories that need to be covered by global performance standards of:

— Safety;
— Reliability;
— Economics.

Since there are typically one or more performance areas in a performance category, a set of global KPIs needs to be established for each performance standard and criteria associated with a given performance area. There are various common global KPIs currently used by the world nuclear industry for measuring and monitoring overall NPP safety and economic performance. Some of those performance criteria and associated KPIs were either developed by the IAEA [V-4 – V-7] for collection and dissemination in the IAEA's Power Reactor Information System (PRIS), or by other industry groups, such as WANO/INPO [V-8] for benchmarking and inspection/review. Some of these global KPIs may also be applicable to KM directly or by derivation and can be used.

There are specific performance areas (including the KM performance areas), that serve the business objectives and strategies and global performance areas, with associated specific performance criteria, against which the KM specific KPIs are established, measured, and monitored. Other KPIs can be established for specific KM performance areas.

As a NPP gets closer to the end of its operating license term, in some cases, the owner/operating organization decides to extend the licensed operation period. In order to make that decision, special safety and economic reviews are conducted for the purpose of determining if a safe and economic longer-term operation (LTO) by renewal/extension of license is feasible.

For example, in technical assessments, specific knowledge and information on ageing mechanisms and their impact particularly on essential plant SSCs and associated actions to manage them in an LTO, become more rigorous for reviews and evaluations toward license renewal/extension. This places unique demands on the availability of competent, qualified, and capable technical human resources. Accordingly, KM plans and processes need to be adjusted to support the LTO activities and the operational activities in the longer term, especially when managing risk of knowledge loss, as well as acquisition, improvement, share and use of new knowledge in, for example, testing or sampling, material ageing, engineering methods, maintenance, surveillance, and inspection, in its KM practices.

Here, it is important to identify and maintain the organization's knowledge, e.g., internal and external knowledge sources, utilization of knowledge, knowledge sharing, and preservation of organizational knowledge and to capture tacit knowledge. Particularly, performances in maintaining KM tools, such as information management as part of knowledge management, or other methods to allow for quick peer and review of ageing issues and gaps in knowledge [V-9], [V-10], need to be monitored, measured, and assessed through KM specific KPIs. These KPIs are additions to the already existing KPIs for measuring and monitoring the performance of KM methods/processes for ensuring the preservation of plant SSCs history and experiences and all relevant design, operation and maintenance data documented and accessible for LTO.

The KM and KM programme KPIs also need to monitor and measure effective collecting, sharing, and preserving research and operational findings and knowledge related to LTO from OPEX.

V-1.9. Operation in Transition to Decommissioning

Before the start of decommissioning activities, there is a period of transition from operations to decommissioning that requires decisions mainly made by the owner/operating organization. Following the decision on ceasing operation and on permanently shutting down for the nuclear facility starts transition period. Between the decision to shut down and the actual end of operations, the owner/operating organization makes decisions on several elements of decommissioning, such as the decommissioning plans and paths for management of large volumes of waste generated by the decommissioning, including the waste collected during the facility operation. Particularly, how the facility is designed, sited, and operated has significant importance to decommissioning goals, plans and activities, including characterization and handling of the waste.

During the transition phase information and knowledge need to be managed, primarily involving identification, collection, and transfer of knowledge from the operating organization to the decommissioning organization.

In general, the owner/operating organizations of nuclear facilities rely on specialized organization for the decommissioning of their facilities. However, the owner/operating organization plays a central role in preparation for decommissioning strategies, programmes, and plans, as it possesses an accumulated and detailed design and operation history of the plant and has the most authentic knowledge of the legal, regulatory, technical, and financial information. Failure to manage this knowledge and experience can lead to additional work, such as waste characterization, etc.

Therefore, it is necessary to develop and implement KM policies and training programmes for the owner/operating organization internal and external staff during the operation to decommissioning transition phase [V-11], as the appropriate use of personnel with knowledge of the facility and its SSCs, as well as design and operation history, is invaluable for successful performance of decommissioning activities. Accordingly, an effective management of knowledge collected by the owner/operating organization ensures that the organization has a long term and coordinated approach with regard to the capture, retention, and update of critical knowledge of the facility towards decommissioning. This approach needs to cover all three key elements of KM, people, process, and tools and the PIs to measure the effectiveness of the KM programme during the transitioning from operation phase to decommissioning phase.

Effective management of knowledge: The experience of the operating organization's staff in managing the facility under operating conditions and their knowledge and familiarity with the features of the facility, are essential for a successful implementation of the decommissioning (for example, as noted in Ref. [V-12], OPEX has shown that personal exposure during decommissioning can be minimized when experienced staff is operating organization and implementing decommissioning).

1. The IAEA Safety Standards [V-13] in Paragraphs 3.4 and 4.4 provide requirements regarding the availability and qualification of personnel involved in a decommissioning project. Therefore, the skills needed for decommissioning need to be evaluated and the minimum requirements for availability and qualifications of staff in each position need to be established and retained.

It is, however, also an observed impact from past transitioning of facilities from operation to decommissioning that the institutional knowledge may be lost due to the departure of qualified and knowledgeable facility staff, which is typically expected to occur after the decision on ceasing operation and on permanently shutting down the nuclear facility is made. Knowledge can also be lost during inappropriate training of new personnel.

Furthermore, as discussed in Ref. [V-14], the need for management of institutional knowledge heavily relies on the time between the decision to shutdown of the facility and the beginning of decommissioning. Naturally, the longer the time between the decision to shutdown of the facility and the beginning of decommissioning takes, the greater this problem adversely affects the execution of decommissioning.

2. Accordingly, the effectiveness of KM (particularly the identification and retention) of people in the operating organization with knowledge necessary or beneficial for decommissioning can be measured.

Effective management of processes and tools: Recording and retention of relevant facility design and operation information from the pre-decommissioning period can be critical for timely and financially successful decommissioning. Discovering that the actual plant layout or configuration does not match the plant layout or configuration in the decommissioning plan would result in stopping the decommissioning activities and potential change of the implementing strategy. Such unexpected event may occur if information on the design and configuration history and potential modifications is not effectively communicated from the operational to decommissioning organization.

The knowledge and documents to be preserved need to be identified as early as possible during the operation of the nuclear facility, so that continuous mechanisms (processes, tools, and methods) for recording and storing of relevant information for use in the decommissioning phase can be established. Information needs to be stored in such a way that integrity can be guaranteed as time goes by and that knowledge can be easily accessed in the future. Accordingly, it is crucial to have a sound management system that provides for the identification, storage, and easy retrieval of both paper and electronic documents.

It is therefore important to consider 'pre-decommissioning' as an operational phase in the lifecycle of a nuclear facility and to preserve during operation the records and information that might be useful after shutdown. To preserve records, discipline and commitment are required, taking into account that some of the information to be recorded does not have an immediate use, which is why this activity might be seen as time consuming and non-productive.

Training has an important role during the transition to decommissioning, when the detailed design of the decommissioning project and its organization are being developed. Training can be an effective tool to transmit information stored during the operation of the facility to the decommissioning organization and its personnel. Training is also essential during the planning and performance of specific decommissioning tasks, particularly during the detailed planning of each work package, which usually relies on a sound knowledge of the operational history of the systems to be dismantled [V-11 – V-13].

V-1.10. Plant Disposal (Decommissioning) Knowledge Management Performance Key Indicators

During the execution of decommissioning plans and project, information and KM affects the successful achievement of decommissioning strategic objectives which focus on removing activated and contaminated materials from the facility safely and efficiently that would include, for example:

- Keeping the amount of radioactive waste to a minimum;
- Maintaining the costs of the decommissioning as low as possible;
- Optimizing the clearance or preservation of structures for further unrestricted reuse;

— Protecting workers, public and environment from unnecessary radiological exposure by measures, such as applying necessary monitoring and surveillance, taking appropriate measures to prevent contamination or dispersion, protect the people and environment from hazards created by the decommissioning process.

REFERENCES

[V-1] INTERNATIONAL ATOMIC ENERGY AGENCY, Milestones in the Development of a National Infrastructure for Nuclear Power, IAEA Nuclear Energy Series No. NG-G-3.1 (Rev. 1), IAEA, Vienna (2015).

[V-2] INTERNATIONAL ORGANISATION FOR STANDARDIZATION, Knowledge Management Systems – Requirements, ISO Standard No. 30401, ISO, Geneva (2018).

[V-3] INTERNATIONAL ATOMIC ENERGY AGENCY, Fundamental Safety Principles, IAEA Safety Standards Series No. SF-1, IAEA, Vienna (2015).

[V-4] INTERNATIONAL ATOMIC ENERGY AGENCY, Operational Safety Performance Indicators for Nuclear Power Plants, IAEA TECDOC Series TECDOC-1141, IAEA, Vienna (2000).

[V-5] INTERNATIONAL ATOMIC ENERGY AGENCY, Economic Performance Indicators for Nuclear Power Plants, IAEA Technical Reports Series No. 437, IAEA, Vienna (2006).

[V-6] SPITZER, D. R., Transforming Performance Measurement: Rethinking the Way We Measure and Drive Organisational Success, HarperCollins Focus, Nashville (2007).

[V-7] INTERNATIONAL ATOMIC ENERGY AGENCY, Power Reactor Information System (PRIS), https://www.iaea.org/pris/

[V-8] WORLD ASSOCIATION OF NUCLEAR OPERATORS, Performance Objectives and Criteria, WANO PO&C 2013-1, WANO, London (2013).

[V-9] INTERNATIONAL ATOMIC ENERGY AGENCY, Plant Life Management Models for Long Term Operation of Nuclear Power Plants, IAEA Nuclear Energy Series No. NP-T-3.18, IAEA, Vienna (2015).

[V-10] INTERNATIONAL ATOMIC ENERGY AGENCY, SALTO Peer Review Guidelines, IAEA Services Series No. 26, IAEA, Vienna (2013).

[V-11] INTERNATIONAL ATOMIC ENERGY AGENCY, Decommissioning of Nuclear Facilities: Training and Human Resource Considerations, IAEA Nuclear Energy Series No. NG-T-2.3, IAEA, Vienna (2008).

[V-12] INTERNATIONAL ATOMIC ENERGY AGENCY, Selection and Use of Performance Indicators in Decommissioning, IAEA Nuclear Energy Series, No. NW-T-2.1, Vienna (2011).

[V-13] INTERNATIONAL ATOMIC ENERGY AGENCY, Decommissioning of Facilities, IAEA Safety Standards Series No. GSR Part 6, IAEA, Vienna (2014).

ANNEX VI.

NATIONAL EXAMPLES OF PERFORMANCE INDICATORS

This Annex provides several examples of PIs from Member States.

VI-1. FRAMATOME, FRANCE

During the initial stage of implementation of KM programme, KPIs were basically project-centric: meeting the deadlines of organization's platform projects (wiki, search, etc.). Further improvement in implementation of the KM programme through elaboration of new policies and processes for KM, resulted in identification of new KPIs, which were compiled in a single "earned value" index. As shown below, KPIs measure the degree to which they are applied in the organization as well as their effectiveness:

0- Scale of five levels of adoption (strength index) was introduced to measure implementation of KM polices and processes. Not applied.
1- Initiated (we start) > New processes and policies are communicated; platforms are known > measuring awareness.
2- Developing (we use) > New processes and tools are used > measuring activity.
3- Confirmed (we produce) > Knowledge objects are produced and stored > measuring output.
4- Performance > (we reuse) > Knowledge objects are leveraged for reuse, thus saving time and reducing risks > measuring outcome.
5- Excellence (We engage; we acknowledge the value) > Employees and customers are engaged > measuring results.

This evaluation is based on a large number of precise KPIs such as:

— Average time spent on a qucry on the search engine;
— Degree to which eLearning modules are viewed until the end;
— Percentage of experts belonging to a community of practice;
— Percentage of knowledge bases complying with company classification standards;
— Percentage of knowledge content with a positive appreciation/comment; etc.

The "strength index" is defined from a compilation of various KPIs. For example, to identify if organization is in stage 3 (confirmed), it is needed to answer on a series of statements such as:

— All critical knowledge domains of the entity have a knowledge base attached;
— All experts less than 2 years away from retiring have a knowledge transfer plan underway;
— All official documents are indexed by the search engine, etc.

These statements need to be validated. Associated values are compared to the associated performance criteria with determined metrics, assessment criteria, rules, and methods for each performance indicator.

VI-2. ONR, UK

Knowledge Management risk analysis status, KM activities, and knowledge transfers need to be reviewed to identify the risks, activities, and good practices. The KM health reporting guidance of the ONR, UK is presented in Table VI-1.

TABLE VI-1. KNOWLEDGE MANAGEMENT HEALTH REPORTING GUIDANCE

Metric	**RAG Criteria / Guidance**
KM Risk Analysis Status This need to include any risks highlighted and date analysis completed/last reviewed.	Red: Unsighted on current knowledge risk. Amber: Aware of the risks however behind plan with mitigation and/or no mitigation in place. Green: Aware of the risks, planned the mitigations and maintaining a focus on them. The risks are clear to the business. Blue: Example of best practice which can be shared with other training committees
KM Activities Evidence of KM organizational learning activities during reporting quarter, such as post-job briefings, learning briefs, task files, etc.	Red: Evidence of missed opportunities to capture and protect critical knowledge and/or no evidence of KM activities being undertaken. Amber: Evidence of missed opportunities to capture and protect knowledge and/or limited evidence of KM activities being undertaken. Green: Evidence of KM activities being undertaken; effective use of available knowledge capture tools, and KM processes being demonstrated. Blue: Example of best practice which can be shared with other training committees.
Knowledge Transfers Evidence of knowledge transfers conducted during reporting quarter.	Red: Evidence of critical knowledge lost with no knowledge transfer activity conducted. Amber: Evidence or risk of knowledge lost and/or planned knowledge transfer activity not completed. Green: Knowledge transfer activities completed to plan (if none planned in quarter still can be green). Blue: Example of best practice which can be shared with other training committee.

VI-3. CEZ, CZECH REPUBLIC

Knowledge Management is an integral part of the CEZ Group organization's safety and environmental protection policy, supported by human resources processes, training and competence development, benchmarking, lecturers, university collaboration and research. This programme focuses on senior knowledge holders who will retire or leave their positions within two years and is monitored with KPIs to measure the effectiveness of KM. CoPs are designated networks of people who share knowledge and lessons learned and are supported in their personalized development through tailor-made training. The KPIs for a CoP are new and are still being improved. The Performance Indicators measure the percentage of knowledge transfer when a knowledge employee leaves the company, Communities of Practice Openness, and Successor's score.

VI-3.1. Key Performance Indicator to Measure the Performance of the Knowledge Management Programme

Introduction of CEZ Group

The main activities of CEZ are the generation, distribution, trading, and sale of electricity and heat; the trading and sale of natural gas; the provision of comprehensive energy services from the new energy sector and coal mining.

VI-3.2. Current Status of Knowledge Management Implementation in CEZ

The systematic approach to KM in CEZ, dates to 2008. Knowledge management is a vital part of CEZ's organizational culture (values, attitudes, and work behaviour of employees) and it is described as one of the five essential work behaviours supported by the company, as shown below:

Principle of Work Behaviour – Expertise: to stay competitive, employees must constantly develop and jointly share their knowledge and skills.

Knowledge management is part of the IMS on the level of processes. The strategy and methodology for KM is determined and regularly updated. KM is supported by human resources processes (e.g., employee adaptation, appraisal interview, training and development, succession planning, exit interview, etc.), training and competence development for KM based on the WANO and IAEA recommendations, documents, and experience. Benchmarking, experienced lecturers, university collaboration and research are also utilized.

CEZ also uses managerial and employee feedback as well as self-assessment tools, to improve methods, procedures, and documentation processes for KM.

The technical platform for KM in CEZ is called the 'Knowledge portal' and it is an integral part of the company's intranet. To capture and share knowledge, a SharePoint library is used as a repository for experience reports, handbooks, and other materials. This material and documents support the training and adaptation processes. To transfer knowledge we use face-to-face tools, such as cooperation in the Communities of Practice (CoPs), doubling of positions, succession planning, etc. To encourage the intergenerational transfer of knowledge in the organization, the KM 'Senior Mentor' programme was created. This programme focuses on senior knowledge holders who will retire or leave their positions within two years where the company needs to be sure that they will transfer as much tacit knowledge and experience as possible. For these Senior Mentors we create special positions in the organization to free up their capacity to transfer their vital knowledge. This programme was launched in 2020 and CEZ still works to improve it so that it fits into the company needs.

VI-3.3. Way of Monitoring Knowledge Management Programme

CEZ monitors KM with a set of KPIs annually, uses the KPIs to measure the activities held in KM and the effectiveness of KM. Management, employee feedback, benchmarking, and self-assessment tools support the KPIs.

VI-3.4. Way of Measuring Effectiveness of Knowledge Management Programme

List of KPIs that are used to monitor and evaluate the effectiveness of the KM programme:

- Number of knowledge workers in total;
- Number of newly nominated knowledge workers;
- Number of newly nominated knowledge workers with a high risk of loss;
- Number of CoPs;
- Number of experts;
- Number of documents uploaded to the knowledge portal;
- Number of experience reports uploaded to the knowledge portal;
- Number of visitors accessing the knowledge portal, library; etc.
- Number of experience reports used in training compared to the total number of experience reports;
- Percentage of realized knowledge transfers (any KM tool used) compared to the number of knowledge employees who left in the calendar year;
- Succession KPI – successors' score;
- Openness of CoPs (%) ratio of participants from other entities to participants in the managing (core) entity.

VI-3.5. Good Practices to Share in this Area

Communities of Practice are designated networks of people who share information and knowledge. They represent KM in real life, and they are one of the most important KM mechanisms. This concept was launched as a pilot in CEZ in 2018. The aim is to share knowledge and lessons learned, develop expert knowledge, and create a network across the departments and topics to use the synergy effect. Today, CEZ has more than 30 active Communities of Practice (CoPs). These communities are encouraged to create handbooks, organize joint meetings, and develop specialized tools such as e-books, online collaboration platforms, shared spaces in the knowledge portal, and a SharePoint library. Personalized development for CoP experts and leaders is supported through tailor-made training programs.

Key performance indicators (KPIs) for CoPs include metrics such as the number of experts involved, the frequency of meetings, and the valuable documents shared. To assess the openness of a CoP to its surroundings, an additional KPI was created. The KPIs for CoPs are relatively new and are currently undergoing further refinement.

VI-3.6. Examples of Knowledge Management Related Performance Indicators

Examples of CEZ KM related PIs are presented in Table VI-2. The PIs measure knowledge transfer when knowledge holders leave the organization, Communities of Practice openness, successor's score, and the internal potential of successors to support knowledge sharing.

TABLE VI-2. CEZ KNOWLEDGE MANAGEMENT RELATED PERFORMANCE INDICATORS

Performance Indicator	**Percentage of Knowledge Transfer When Knowledge Holder Departs**	**Communities of Practice Openness**	**Successor's Score**
Definition	Data Source: HR reports, KM reports. Data requirements: number of knowledge holders who left the company in a calendar year, number of knowledge transfers in a calendar year. Calculation: Number of realized knowledge transfers to the number of key knowledge holders who left the company in a calendar year (%).	Data Source: HR reports, KM reports. Data requirements: Number of Community of Practice (CoP) meetings, number of participants in the CoP meetings (internal and external in detail of organizations and departments). Calculation: Percentage ratio of participants from other entities compared to participants in the managing (core) entity at joint meetings.	Data Source: HR reports, KM reports. Data requirements: Number of successors, personal changes within the calendar year in total (succession positions), number of promoted successors. Calculation: Number of successors who succeeded for new position compared to the number of succession positions where any personal change was done in the calendar year (%).
Identify a basis for the new/changed indicator (i.e., why the new indicator is needed or why it is changed)	The KPIs check whether there were knowledge holders leaving without knowledge transfer. Used KM tools for transfers are tracked in detail.	To assess how open the CoP is to support the knowledge sharing compared to its surroundings.	The KPI checks whether the succession pool is the main source of relevant candidates for filling the open key positions.

VI-4. ROSATOM, RUSSIAN FEDERATION

The main objectives of the KPIs (Key Performance Indicators) for the NKM (Nuclear Knowledge Management) process are the development and improvement of the system itself, ensuring the preservation of critical knowledge, as well as the ability to manage this process by monitoring its development and the degree of its maturity.

This section provides examples of KPIs developed for the design organization and two examples of KPIs developed by the utility which is the owner of nuclear facilities in the Russian Federation.

VI-4.1. Rosatom the Design Organization

The indicators presented below allow a reader to understand how NKM processes are developed and changed. To appropriately consider the dynamics of the KM programme development it is important to include KPI maps (a KPI map is a set of indicators, which allows top managers to evaluate the performance and effectiveness and to measure the degree of achievement of its objectives).

When forming a KPI map, it is possible to adjust the degree of importance of a particular task by changing its value (from 0 to 100 percent, and, accordingly, its impact on the process, its effectiveness).

The types of indicators placed in the KPI maps may differ for different top managers as well as their contribution to complex KPIs.

The presented examples of indicators allow readers to combine individual indicators into the complex KPIs required at this stage, which need to motivate the development of the NKM system.

To properly understand the process of managing KPIs, it is useful to determine the types of KPI including their classification.

The proposed types of KPIs, as well as the options for their use, are not final and can be further developed and adjusted. The figure below shows the proposed classification of KPIs.

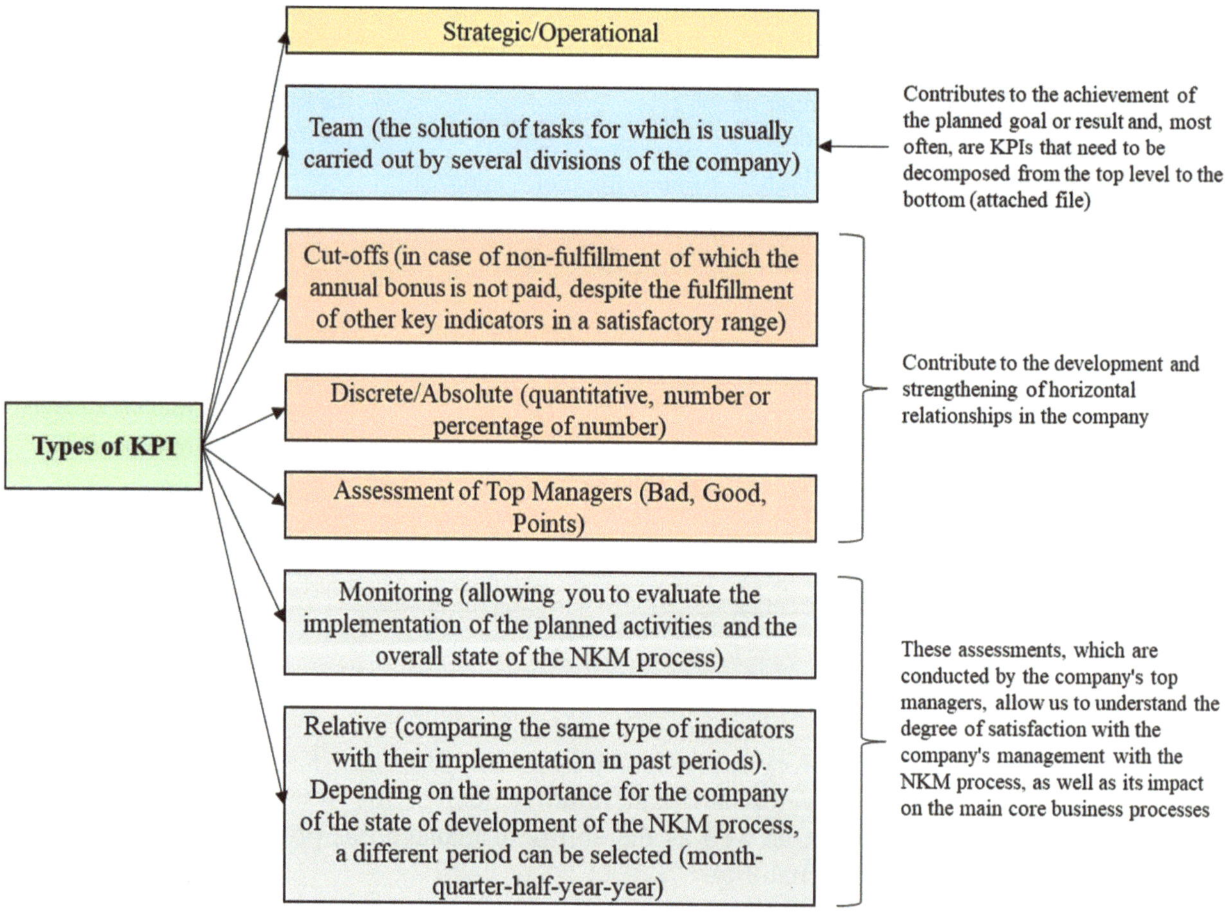

FIG. VI-2 Classification of the KPIs.

TABLE VI-3. EXAMPLES OF PARAMETERS THAT CAN BE USED IN THE FORMATION OF COMPLEX KPIs

No.	Name	Value
1	Availability of the NKM Policy	Yes/No-1/0
2	Availability of the NKM Business Process (with description)	Yes/No-1/0
3	Development of NKM goals for the current year, as well as for a longer-term period (3-5 years)	Yes/No-1/0
4	Assessment of the level of development of the process NKM (embedding to an integrated enterprise management system/part of the maturity process NKM), including a detailed description of processes and sub-processes NKM	Target Value - 1
5	Availability of a Process Development Plan for the current year to achieve the set goals, as well as a Plan for the period of 3-5 years	Yes/No-1/0
6	Availability of the NKM development programme	Yes/No-1/0
7	Availability of the system / database / portal/a site for storing and distributing and accessing KVZ to young people	Yes/No-1/0
8	Appointment of a TOP manager for NKM (order of appointment, inclusion in her/his job description of the requirements for NKM, familiarization with it)	Yes/No-1/0
9	Distribution of responsibilities for the NKM process between divisions (orders, orders, procedures, inclusion of requirements for the NKM process in the regulations on divisions, appointment of responsible persons in divisions, familiarization)	Yes/No-1/0
10	Completeness of the NKM business process description Availability of a document included in the IMS, describing all the processes involved in the NKM activity (identification, extraction, preservation, organization of access, distribution, etc.)	Target Value - 1
11	Number of specialties covered by the NKM process / to the total number of project specialties of the company	
12	Number of experts involved in the NKM process/ total number of employees 60+	
13	Number of completed NKM programme items/ total number of items that have expired.	
14	The number of identified carriers of critical knowledge / to the total number of employees of the enterprise 60+	
15	Number of identified critical knowledge/ number of documented critical knowledge	
16	Manager's rating (likes/dislikes/point)	
Comparative data		
17	Number of documented critical knowledge for Q_1 (first half/first year)/ Q_2 (second half/second year)	
18	Number of employees who carry critical knowledge/number of critical knowledge transfer programs (Mentoring, Coaching)	
19	Identification, analysis and assessment of NKM risks. Availability of a risk management programme	

For assessment of the level of development of the process NKM, including a detailed description of processes and sub-processes NKM (Point 4 in the Table 4-3) the following formula is used:

$$N_{dev} = \frac{N_1 \cdot N_2}{N_1}, \qquad \text{(VI.1)}$$

where N_1 – total number of documents in the Integrated Management System (IMS);

N_2 – the number of business processes of the enterprise;

N_3 – the number of documents for NKM processes.

The completeness of the NKM business process description is calculated by the following formula:

$$N_{compl} = \frac{N_4}{N_5}, \qquad \text{(VI.2)}$$

where N_4 – the number of described NKM processes;

N_5 – total number of processes in the system NKM.

VI-4.2. Rosenergoatom

Examples of two KM related Performance Indicators developed at the utility level.

TABLE VI-4. DETAILS ON KM RELATED: COMMISSIONING ABNORMALS INDICATOR

Data	**Description**
KPI ID	$F_{3.1}$ – a ration of abnormalities/violations happened at a new build unit due to Commissioning and Operations Management System weaknesses, per year
KPI Owner	Deputy Director General on Operations
Process Owner	Director, Division on new build operational readiness
Frequency to review	Once per month
Initial data	A trigger is any of malfunction/violation due Commissioning phase weaknesses
Target level	0.022
Anomaly level	0.0275
Actions to response	>0.022

Comments: $F_{3.1}$ KPI demonstrates how well Operations Management handles with knowledge during commissioning stage.

This knowledge includes but not limited to:

- specifics of works to be done was known in detail:
- operational expenditure of the commissioning previously done at another construction units were considered as appropriate;
- database was used to preparing/planning commissioning operations;
- competence of participants involved in the process was appropriate;
- quality of documents is due knowledge was incorporated before use;
- others.

For calculation the indicator demonstrated how well Operations Management handles with knowledge during commissioning stage, the following formula is recommended to use:

$$F_{3.1} = \frac{RC_{com}}{RC_{total}}, \qquad \text{(VI.3)}$$

where RC_{com} – a number of root causes defined at licensed new build unit operations due to commissioning and Ops management weaknesses as per Event Investigation Reports (EIRs);

RC_{total} – a general number of all root causes defined at licensed new build unit operations as per all EIRs.

TABLE VI-5. DETAILS ON KM RELATED: REPEATED ABNORMAL EVENTS INDICATOR

Data	**Description**
KPI ID	K_{re} – a ratio of repeated abnormal events happened at operational power units, per year
KPI Owner	Director, Division on Safety and Production Control
Process Owner	Deputy Director General on Operations
Frequency to review	Weekly: KPIs monitoring for emerge actions as appropriate. Yearly: provide analyse and evaluation of effectiveness of the proactive actions done
Initial data	Event Investigation Reports; NPP reports; OPEX DBs
Actions to response	A trigger is a deviation from the target level or negative trends
Comments 1: The current KPI level is 0,086. The ambitious goal is to decrease a number of repeated events to K_{re}<0,05	
Comments 2: K_{re} KPI demonstrates how effective knowledge (lessons learned) is used to coupe the issue (process, human, technology weaknesses). This knowledge includes but is not limited to: — Knowledge (lessons learned) was used to improve processes; — OPEX from similar plants was used to develop countermeasures; — New knowledge gained from the events has been properly kept in the plant manuals/procedures and training materials; — there is evidence that repeated events would not be possible. — others.	

For the related abnormal events indicator, the value of 0.05 is considered as a “target level” and the value of 0.086 as “anomaly level”. This indicator could be calculated by the following formula:

$$K_{re} = \frac{N_{re}}{N_{gen}}, \quad \text{(VI.4)}$$

where N_{re} – a number of repeated events at licensed operational power units, events/year;

N_{gen} – a general number of all events at all licensed operational power units, events/year.

GLOSSARY

A more comprehensive list of useful terms and definitions can be found on the NKM Hub (https://nucleus.iaea.org/sites/connect-members/nkmh/Pages/NE-Glossary.aspx). The IAEA Safety Glossary need to be regarded as the formal reference for the glossary of terms from the Agency.

critical knowledge. Critical knowledge is knowledge that is considered as the most significant knowledge whose loss, or failure to maintain and implement correctly, carries a direct and immediate high-risk of safety or commercial loss.

communities of practice. A voluntary group of peer practitioners who share lessons learned, methods, and best practices in a given discipline, or for specialized work. The term also refers to a network of people who work on similar processes or in similar disciplines and who come together to develop and share their knowledge in that field for the benefit of both themselves, and their and other organization(s).

concept sorting and mapping. Tools for organizing and representing knowledge. Concept maps include concepts, usually depicted in circles or boxes of some type and relationships between concepts or propositions, indicated by a connecting line between two concepts.

effectiveness. Extent to which planned activities are realized and planned results achieved.

e-learning. An abbreviation of electronic learning, which uses information technology systems to conduct education or training as well as to manage those related activities. Services that are delivered, enabled, or mediated by information and computer technologies for the purposes of conducting education or training and the technology and services that help create, manage, and deliver those activities.

enterprise resource planning. Integrated management of main business processes, often in real time and mediated by software and technology.

expert/intellectual systems. A knowledge-based system that provides for solving problems in a particular domain or application area by drawing inferences from a knowledge base developed from human expertise.

explicit knowledge. Knowledge that has been articulated or has already been codified in some form such as manuals, procedures, databases, or electronic media. Knowledge that can be easily expressed in documents. *See also* ***knowledge***.

implicit knowledge. Knowledge that is held in a person's mind and has not yet been captured or transferred in any form. *See also* ***knowledge***.

institutional knowledge. The collective knowledge of all the employees working in an organization or institution.

integrated management system. A single coherent management system for facilities and activities in which all the component parts of an organization are integrated to enable the organization's objectives to be achieved.

knowledge. Knowledge is a mix of experiences, values, contextual information, and expert insight for acquiring, understanding, and interpreting information. Together with attitudes and skills, it forms a capacity for effective actions. Note: knowledge is a combination of 'knowing facts' about something and 'knowing how' to do something. It refers to a body of facts and principles accumulated by humankind over the course of time. It is distinct from information, as knowledge is information that has a purpose or use. Data lead to information and information leads to knowledge. Knowledge confers a capacity for effective action.

knowledge capture. A process of capturing the knowledge available within an organization and making it available.

knowledge management. An integrated, systematic approach to identifying, managing, and sharing an organization's knowledge and enabling groups of people to create new knowledge collectively to help achieve the organization's objectives.

knowledge management databases. A collection of information organized in such a way that a computer program can quickly select desired pieces of data. Relational databases are organized by fields, records, and tables. A field is a single piece of information, a record is one complete set of fields, and a table is a collection of records. Storing content in fields rather than on static pages makes that content appropriate for dynamic delivery.

knowledge map. Overview of knowledge assets in an organization. It shows the distribution and correlation of knowledge as well as providing navigation for potential users so that they can find the desired knowledge properly. Graphical (diagrammatic) techniques to show associations, linkages, structure and inter-relationships in concepts or knowledge domains. Note: knowledge mapping is a process to determine where knowledge assets are in an organization and how knowledge flows operate within the organization. Evaluating relationships between holders of knowledge will then illustrate the sources, flows, limitations, and losses of knowledge that can be expected to occur. See also **concept sorting and mapping**.

knowledge management system. A knowledge management system (KMS) is an IT infrastructure designed for supporting the KM process by tools and technologies for management of knowledge and information.

knowledge management programme. The process of organization intending to implement knowledge management principles, practices, and projects in order to achieve organizational objectives with the intention of improving an organization's performance.

knowledge loss risk assessment. Method and process of assessing critical knowledge and resources in an organization and the risk of them being lost due to attrition. A process used to determine the potential business impact of the loss of critical knowledge from an organization.

knowledge loss risk management methodology. The goal of the knowledge loss risk management (KLRM) methodology in nuclear organizations is to provide assurance that knowledge preservation & knowledge transfer programmes are properly considered throughout the different phases of a nuclear project.

key performance indicators. A set of quantitative & qualitative measurements used to gauge an organizations overall performance.

lessons learned. Concise descriptions of knowledge derived from experiences that can be communicated through mechanisms such as storytelling, debriefing etc, or summarized in databases.

management system. A set of interrelated or interacting elements (*system*) for establishing policies and objectives and enabling the objectives to be achieved in an efficient and effective manner.

performance indicator. A metric that provides a measurement of the activities or processes that underpin a business objective.

portal/intranet. A tool to integrate many existing systems within an organization as well as to provide a solid platform to develop other knowledge management initiatives, enhancing the efficiency of communication and of organizational processes. A portal that greatly facilitates the production, accessibility, sharing and effective use of valuable information. It also guarantees generation and usage of information at different times or across different locations and teams.

preparatory activities. Activities that establish planning or preparation for the conduct of the technical support, such as the determination of tasks, scopes, schedules, competencies for technical support needed (or to be provided) and technical input needed (or to be provided) — for example, what data to collect/monitor and how to collect/monitor them, or the reason, basis or justification for needed technical support (such as identification of the issue(s)).

simulation tools. Learning tools in which a real-life situation is simulated using models or interactive computer programmes.

standards of performance. Specified goals, criteria, thresholds that are expected to be met to ensure effective achievement of business/corporate/organization strategic objectives.

storytelling. The practice of relating personal recollections, impressions, perspectives, observations, and interpretations, typically with the aim of sharing a particular series of events that collectively convey a message that is of use to the listeners.

search engine. A mechanism that identifies which items, out of a given collection, conform to a given query string.

tacit knowledge. Tacit knowledge is knowledge that is wholly embodied in the individual, is rooted in practice, experience, intuition, and individual skills and is difficult or even impossible to recall, articulate and thus to transfer. Note: tacit knowledge is the knowledge held in the mind of individuals and is often unspoken and difficult to articulate, share, or transfer. However, it may be partially transferred from one individual to another individual using different tools and methods. The consensus amongst knowledge management professionals is that most of the knowledge in any organization is tacit.

LIST OF ABBREVIATIONS

APA	activity performance area
APC	activity performance goal
APG	activity performance criterion
CoP	community of practice
ERO	emergency response organisation
ERT	emergency response team
ES	expert system
GPA	global performance area
GPC	global performance criterion
GPG	global performance goal
HRD	human resource development
I&C	instrumentation and control
IMS	integrated management system
INPO	Institute of Nuclear Power Operators
INSAG	International Nuclear Safety Advisory Group
KM	knowledge management
KMAV	knowledge management assist visit
KPI	key performance indicator
LTO	long term operation
NEPIO	nuclear energy programme implementing organisation
NKM	nuclear knowledge management
NPP	nuclear power plant
OJP	on the job participation
OJT	on the job training
OPEX	operating experience
PC	performance criterion
PI	performance indicator
PI&R	problem identification and resolution
PRIS	Power Reactor Information System
R&D	research and development
ROI	return on investment
SKPI	specific key performance indicator
SMART	specific, measurable, achievable, relevant, time-bound
SO	strategic objective

SPA	specific performance area
SPC	specific performance criterion
SPG	specific performance goal
SSC	system, structure, and component
WANO	World Association of Nuclear Operators

CONTRIBUTORS TO DRAFTING AND REVIEW

Abbasova, D.	Institute of Resource Ecology, HZDR Germany
Bilic Zabric, T.	International Atomic Energy Agency
Cvetkov, I.	Kozloduy Nuclear Power Plant, Bulgaria
Drace, M.	International Atomic Energy Agency
Drury, D.	International Atomic Energy Agency
Dugage, M. R.	Areva, France
Gordon, I.	International Atomic Energy Agency
Ganesan, A.	International Atomic Energy Agency
Ivanov, D.	JSC EC ASE, Russian Federation
Kang, K.	Independent consultant, Republic of Korea
Kilic, N.	NESiS, LLC, United States of America
Lemke, M.	INPO, United States of America
Malik, H.	Information Security and Governance, United Kingdom
Kvetonova, R.	International Atomic Energy Agency
Ovanes, M.	International Atomic Energy Agency
Page, S.	EDF Energy, United Kingdom
Parkhimenka, U.	BSUIR, Belarus
Piciaccia, L.	Norwegian Radiation and Nuclear Safety Authority, Norway
Roberts, J.	International Atomic Energy Agency
Ruzickova, M.	ČEZ a.s., Czech Republic
Semenova, D.	JSC Rosatom, Russian Federation
Simic, Z.	EC/JRC, The Netherlands
Shi, Q.	WANO, United Kingdom
Smith, E.	Independent consultant, United Kingdom
Yuzhakov, A.	VNIIAES, JSC, Russian Federation
Zhivitskaya, H.	International Atomic Energy Agency

Technical Meeting

Vienna, Austria: 29 September - 2 October 2020

Consultants Meetings

Vienna, Austria: 03-06 March 2020, 13-16 April 2021, 24-28 January 2022

IAEA
International Atomic Energy Agency

CONTACT IAEA PUBLISHING

Feedback on IAEA publications may be given via the on-line form available at:
www.iaea.org/publications/feedback

This form may also be used to report safety issues or environmental queries concerning IAEA publications.

Alternatively, contact IAEA Publishing:

Publishing Section
International Atomic Energy Agency
Vienna International Centre, PO Box 100, 1400 Vienna, Austria
Telephone: +43 1 2600 22529 or 22530
Email: sales.publications@iaea.org
www.iaea.org/publications

Priced and unpriced IAEA publications may be ordered directly from the IAEA.

ORDERING LOCALLY

Priced IAEA publications may be purchased from regional distributors and from major local booksellers.

25-00440E

FSC
www.fsc.org
MIX
Papier | Fördert gute Waldnutzung
FSC® C083411

Zeitfracht Medien GmbH
Ferdinand-Jühlke-Straße 7
99095 Erfurt, Deutschland
produktsicherheit@kolibri360.de